UCLA Symposia on Molecular and Cellular Biology, New Series

Series Editor, C. Fred Fox

RECENT TITLES

Volume 108
Acute Lymphoblastic Leukemia, Robert Peter Gale and Dieter Hoelzer, *Editors*

Volume 109
Frontiers of NMR in Molecular Biology, David Live, Ian M. Armitage, and Dinshaw J. Patel, *Editors*

Volume 110
Protein and Pharmaceutical Engineering, Charles S. Craik, Robert J. Fletterick, C. Robert Matthews, and James A. Wells, *Editors*

Volume 111
Glycobiology, Joseph K. Welply and Ernest Jaworski, *Editors*

Volume 112
New Directions in Biological Control: Alternatives for Suppressing Agricultural Pests and Diseases, Ralph R. Baker and Peter E. Dunn, *Editors*

Volume 113
Immunogenicity, Charles A. Janeway, Jr., Jonathan Sprent, and Eli Sercarz, *Editors*

Volume 114
Genetic Mechanisms in Carcinogenesis and Tumor Progression, Curtis Harris and Lance A. Liotta, *Editors*

Volume 115
Growth Regulation of Cancer II, Marc E. Lippman and Robert B. Dickson, *Editors*

Volume 116
Transgenic Models in Medicine and Agriculture, Robert B. Church, *Editor*

Volume 117
Early Embryo Development and Paracrine Relationships, Susan Heyner and Lynn M. Wiley, *Editors*

Volume 118
Cellular and Molecular Biology of Normal and Abnormal Erythroid Membranes, Carl M. Cohen and Jiri Palek, *Editors*

Volume 119
Human Retroviruses, Jerome E. Groopman, Irvin S.Y. Chen, Myron Essex, and Robin A. Weiss, *Editors*

Volume 120
Hematopoiesis, Steven C. Clark and David W. Golde, *Editors*

Volume 121
Defense Molecules, John J. Marchalonis and Carol L. Reinisch, *Editors*

Volume 122
Molecular Evolution, Michael T. Clegg and Stephen J. O'Brien, *Editors*

Volume 123
Molecular Biology of Aging, Caleb E. Finch and Thomas E. Johnson, *Editors*

Volume 124
Papillomaviruses, Peter M. Howley and Thomas R. Broker, *Editors*

Volume 125
Developmental Biology, Eric H. Davidson, Joan V. Ruderman, and James W. Posakony, *Editors*

Volume 126
Biotechnology and Human Genetic Predisposition to Disease, Charles R. Cantor, C. Thomas Caskey, Leroy E. Hood, Daphne Kamely, and Gilbert S. Omenn, *Editors*

Volume 127
Molecular Mechanisms in DNA Replication and Recombination, Charles C. Richardson and I. Robert Lehman, *Editors*

Volume 128
Nucleic Acid Methylation, Gary A. Clawson, Dawn B. Willis, Arthur Weissbach, and Peter A. Jones, *Editors*

Volume 129
Plant Gene Transfer, Christopher J. Lamb and Roger N. Beachy, *Editors*

Volume 130
Parasites: Molecular Biology, Drug and Vaccine Design, Nina M. Agabian and Anthony Cerami, *Editors*

Volume 131
Molecular Biology of the Cardiovascular System, Robert Roberts and Joseph F. Sambrook, *Editors*

Volume 132
Obesity: Towards a Molecular Approach, George A. Bray, Daniel Ricquier, and Bruce M. Spiegelman, *Editors*

Volume 133
Structural and Organizational Aspects of Metabolic Regulation, Paul A. Srere, Mary Ellen Jones, and Christopher K. Mathews, *Editors*

Please contact the publisher for information about previous titles in this series.

UCLA Symposia Board

C. Fred Fox, Ph.D., Director
Professor of Microbiology, University of California, Los Angeles

Charles J. Arntzen, Ph.D.
Director, Plant Science and Microbiology
E.I. du Pont de Nemours and Company

Floyd E. Bloom, M.D.
Director, Preclinical Neurosciences/
Endocrinology
Scripps Clinic and Research Institute

Ralph A. Bradshaw, Ph.D.
Chairman, Department of Biological
Chemistry
University of California, Irvine

Francis J. Bullock, M.D.
Vice President, Research
Schering Corporation

Ronald E. Cape, Ph.D., M.B.A.
Chairman
Cetus Corporation

Ralph E. Christoffersen, Ph.D.
Executive Director of Biotechnology
Upjohn Company

John Cole, Ph.D.
Vice President of Research
and Development
Triton Biosciences

Pedro Cuatrecasas, M.D.
Vice President of Research
Glaxo, Inc.

Mark M. Davis, Ph.D.
Department of Medical Microbiology
Stanford University

J. Eugene Fox, Ph.D.
Vice President, Research
and Development
Miles Laboratories

J. Lawrence Fox, Ph.D.
Vice President, Biotechnology Research
Abbott Laboratories

L. Patrick Gage, Ph.D.
Director of Exploratory Research
Hoffmann-La Roche, Inc.

Gideon Goldstein, M.D., Ph.D.
Vice President, Immunology
Ortho Pharmaceutical Corp.

Ernest G. Jaworski, Ph.D.
Director of Biological Sciences
Monsanto Corp.

Irving S. Johnson, Ph.D.
Vice President of Research
Lilly Research Laboratories

Paul A. Marks, M.D.
President
Sloan-Kettering Memorial Institute

David W. Martin, Jr., M.D.
Vice President of Research
Genentech, Inc.

Hugh O. McDevitt, M.D.
Professor of Medical Microbiology
Stanford University School of Medicine

Dale L. Oxender, Ph.D.
Director, Center for Molecular Genetics
University of Michigan

Mark L. Pearson, Ph.D.
Director of Molecular Biology
E.I. du Pont de Nemours and Company

George Poste, Ph.D.
Vice President and Director of Research and
Development
Smith, Kline and French Laboratories

William Rutter, Ph.D.
Director, Hormone Research Institute
University of California, San Francisco

George A. Somkuti, Ph.D.
Eastern Regional Research Center
USDA-ARS

Donald F. Steiner, M.D.
Professor of Biochemistry
University of Chicago

UCLA Symposia Board membership at the time of the meeting is indicated on the above list.

Glycobiology

Glycobiology

Proceedings of a Smith Kline & French Laboratories-UCLA Symposium on Glycobiology held at Frisco, Colorado, January 14-20, 1989

Editors

Joseph K. Welply
Ernest Jaworski
Department of Biological Sciences
Monsanto Company
St. Louis, Missouri

A JOHN WILEY & SONS, INC., PUBLICATION
New York • Chichester • Brisbane • Toronto • Singapore

Address all Inquiries to the Publisher
Wiley-Liss, Inc., 41 East 11th Street, New York, NY 10003

Copyright © 1990 Wiley-Liss, Inc.

Printed in United States of America

Under the conditions stated below the owner of copyright for this book hereby grants permission to users to make photocopy reproductions of any part or all of its contents for personal or internal organizational use, or for personal or internal use of specific clients. This consent is given on the condition that the copier pay the stated per-copy fee through the Copyright Clearance Center, Incorporated, 27 Congress Street, Salem, MA 01970, as listed in the most current issue of "Permissions to Photocopy" (Publisher's Fee List, distributed by CCC, Inc.), for copying beyond that permitted by sections 107 or 108 of the US Copyright Law. This consent does not extend to other kinds of copying, such as copying for general distribution, for advertising or promotional purposes, for creating new collective works, or for resale.

The publication of this volume was facilitated by the authors and editors who submitted the text in a form suitable for direct reproduction without subsequent editing or proofreading by the publisher.

Library of Congress Cataloging-in-Publication Data

```
UCLA Symposium on Glycobiology (1989 : Frisco, Colo.)
    Glycobiology : proceedings of a UCLA Symposium held at Frisco,
Colorado, January 14-20, 1989 / editors, Joseph K. Welply, Ernest
Jaworski
      p.   cm. -- (UCLA symposia on molecular and cellular biology ;
new ser., v. 111)
    Based on the proceedings of the 1989 UCLA Symposium on
Glycobiology.
    Includes bibliographical references.
    ISBN 0-471-56749-3
    1. Glycoproteins--Congresses.  2. Glycolipids--Congresses.
I. Welply, Joseph K.  II. Jaworski, Ernest.  III. University of
California, Los Angeles.  IV. Title.  V. Series.
    [DNLM: 1. Glyconjugates--congresses.  2. Glycoproteins-
-congresses.  3. Glycosylation--congresses.  W3 U17N new ser. v.
111 / QU 55 U17 1989g]
QP552.G59U25  1989
574.19'2482--dc20
DNLM/DLC
for Library of Congress                                    89-70633
                                                               CIP
```

Contents

Contributors.. ix

Preface
    Joseph K. Welply and Ernest Jaworski......................... xiii

I. GLYCOLIPID ANCHORS

The Glycoinositol Phospholipid Anchor of Human Erythrocyte Acetylcholinesterase
    Terrone L. Rosenberry, William L. Roberts, Jean-Pierre Toutant, Sitthivet Santikarn, and Vernon N. Reinhold...................... 1

The Cytoplasmic Extension as a Determinant for Glycoinositolphospholipid Anchor Substitution
    M. Edward Medof and Mark L. Tykocinski....................... 17

Survival of *Leishmania* Parasites Within Phagocytic Cells: Requirement for Lipophosphoglycan
    S.J. Turco, D.L. McLean, and T.B. McNeely..................... 23

Post-Translational Modifications of the Cellular and Scrapie Prion Proteins
    Stanley B. Prusiner... 35

II. CELL-CELL INTERACTIONS

Sperm Receptor Oligosaccharides as Mediators of Sperm-Egg Interactions in Mice
    Jeffrey D. Bleil and Paul M. Wassarman........................ 59

A Lymphocyte Homing Receptor Is a Lectin, a Member of the Emerging LEC-CAM Family
    S.D. Rosen, J.S. Geoffroy, L.A. Lasky, M.S. Singer, S. Stachel, L.M. Stoolman, D.D. True, and T.A. Yednock........................ 69

Lymphocyte Homing Receptors and Vascular Addressins
    Ellen Lakey Berg, Leslie A. Goldstein, Louis J. Picker, Philip R. Streeter, Mark A. Jutila, Robert F. Bargatze, David F.H. Zhou, and Eugene C. Butcher... 91

The Relationship Between Golgi and Cell Surface $\beta 1, 4$ Galactosyltransferase
    Adel Youakim and Barry D. Shur............................... 107

III. PROTEIN GLYCOSYLATION

Altered Glycosylation of Human Chorionic Gonadotropin in Trophoblastic Diseases and Its Use for the Diagnosis of Choriocarcinoma
Akira Kobata, Junko Amano, Tsuguo Mizuochi, and Tamao Endo...... 115

N-Glycan Processing and the Enterocytic Differentiation of HT-29 Cells Are Related Events
Germain Trugnan, Eric Ogier-Denis, C. Bauvy, M. Aubery, I. Chantret, C. Sapin, P. Codogno, and A. Zweibaum...... 127

Nuclear Pore Glycoproteins: Structure and Function
John A. Hanover, Min Kyun Park, Mara D'Onofrio, Christopher Starr, Tracy S. Olson, and Barbara Wolff...... 145

IV. GLYCOCONJUGATE THERAPEUTICS

Selection of an Expression Host for Human Glucocerebrosidase: Importance of Host Cell Glycosylation
Michel L.E. Bergh, Carol Naranjo, Anita F. Mentzer, Gary D. Barsomian, C. William Christopher, Catherine Bartlett, Shirish Hirani, and James R. Rasmussen...... 159

Glycoconjugates as Drugs
Russell Greig and George Poste...... 173

V. EXTENDED ABSTRACTS OF MEETING-RELATED ARTICLES

These abstracts describe research presented at the meeting by poster participants and plenary lecturers who elected to submit their work for publication in *Journal of Cellular Biochemistry*.

Role of Galaptin in Ovarian Carcinoma Adhesion to Extracellular Matrix *In Vitro*[1]
Howard J. Allen, Daniel Sucato, Barbara Woynarowska, Sally Gottstine, Ashu Sharma, and Ralph J. Bernacki...... 195

Analysis of N-Glycosylation Mutants in *Dictyostelium discoideum*[1]
H.H. Freeze, P. Koza-Taylor, J.A. Jones, and W.F. Loomis...... 197

The Role of Glycosylation in the Transport of Recombinant Glycoproteins Through the Secretory Pathway of Lepidopteran Insect Cells[1]
Donald L. Jarvis, Christian Oker-Blom, and Max D. Summers...... 199

Index...... 201

---

[1] Journal of Cellular Biochemistry (1989/1990), in press. Wiley-Liss, Inc., New York

Contributors

Howard J. Allen, Department of Surgical Oncology, Roswell Park Memorial Institute, Buffalo, NY 14263 [195]

Junko Amano, Department of Biochemistry, Institute of Medical Science, The Univerisity of Tokyo, Tokyo 108, Japan [115]

M. Aubery, INSERM U180, 75006 Paris, France [127]

Robert F. Bargatze, Department of Pathology, Stanford University, Stanford, CA 94305 [91]

Gary D. Barsomian, Department of Microbiology, Genzyme Corporation, Boston, MA 02111 [159]

Catherine Bartlett, Department of Biochemistry, Genzyme Corporation, Boston, MA 02111 [159]

C. Bauvy, INSERM U180, 75006 Paris, France [127]

Ellen Lakey Berg, Department of Pathology, Stanford University, Stanford, CA 94305 [91]

Michel L. E. Bergh, Department of Cell Biology, Genzyme Corporation, Boston, MA 02111 [159]

Ralph J. Bernacki, Department of Experimental Therapeutics, Roswell Park Memorial Institute, Buffalo, NY 14263 [195]

Jeffery D. Bleil, Department of Cell and Developmental Biology, Roche Institute of Molecular Biology, Roche Research Center, Nutley, NJ 07110 [59]

Eugene C. Butcher, Department of Pathology, Stanford University, Stanford, CA 94305 [91]

I. Chantret, INSERM U178, 94807 Villejuif-Cedex, France [127]

C. William Christopher, Process Development, Genzyme Corporation, Boston, MA 02111 [159]

P. Codogno, INSERM U180, 75006 Paris, France [127]

Mara D'Onofrio, Laboratory of Biochemistry and Metabolism, NIDDK, National Institutes of Health, Bethesda, MD 20892 [145]

The numbers in brackets are the opening page numbers of the contributors' articles.

Contributors

Tamao Endo, Department of Biochemistry, Institute of Medical Science, The University of Tokyo, Tokyo 108, Japan **[115]**

H. H. Freeze, La Jolla Cancer Research Foundation, La Jolla, CA 92037 **[197]**

J. S. Geoffroy, Department of Anatomy, University of California, San Francisco, CA 94143-0452 **[69]**

Leslie A. Goldstein, Department of Pathology, Stanford University, Stanford, CA 94305 **[91]**

Sally Gottstine, Department of Surgical Oncology, Roswell Park Memorial Institute, Buffalo, NY 14263 **[195]**

Russell Greig, Department of Cell Sciences, Smith Kline and French Laboratories, King of Prussia, PA 19406 **[173]**

John A. Hanover, Laboratory of Biochemistry and Metabolism, NIDDK, National Institutes of Health, Bethesda, MD 20892 **[145]**

Shirish Hirani, Department of Biochemistry, Genzyme Corporation, Boston, MA 02111 **[159]**

Donald L. Jarvis, Department of Entomology and Institute of Biosciences and Technology, Texas A & M University, College Station, TX 77843 **[199]**

Ernest Jaworski, Department of Biological Sciences, Monsanto Company, St. Louis, MO 63198 **[xiii]**

J. A. Jones, La Jolla Cancer Research Foundation, La Jolla, CA 92037 **[197]**

Mark A. Jutila, Department of Pathology, Stanford University, Stanford, CA 94305 **[91]**

Akira Kobata, Department of Biochemistry, Institute of Medical Science, The University of Tokyo, Tokyo 108, Japan **[115]**

P. Koza-Taylor, La Jolla Cancer Research Foundation, La Jolla, CA 92037 **[197]**

L. A. Lasky, Department of Cardiovascular Research, Genentech, Inc., San Francisco, CA 94080 **[69]**

W. F. Loomis, La Jolla Cancer Research Foundation, La Jolla, CA 92037 **[197]**

D. L. McLean, Department of Biochemistry, University of Kentucky College of Medicine, Lexington, KY 40536 **[23]**

T. B. McNeely, Department of Biochemistry, University of Kentucky College of Medicine, Lexington, KY 40536 **[23]**

M. Edward Medof, Institute of Pathology, Case Western Reserve University, Cleveland, OH 44106 **[17]**

Anita F. Mentzer, Department of Cell Biology, Genzyme Corporation, Boston, MA 02111 **[159]**

Tsuguo Mizuochi, Department of Biochemistry, Institute of Medical Science, The University of Tokyo, Tokyo 108, Japan **[115]**

Carol Naranjo, Department of Cell Biology, Genzyme Corporation, Boston, MA 02111 **[159]**

Eric Ogier-Denis, INSERM U180, 75006 Paris, France **[127]**

Christian Oker-Blom, Department of Entomology and Institute of Biosciences and Technology, Texas A & M University, College Station, Texas 77843; present address: Labsystems OY, 00880 Helsinki, Finland **[199]**

Tracy S. Olson, Laboratory of Biochemistry and Metabolism, NIDDK, National Institutes of Health, Bethesda, MD 20892 **[145]**

Min Kyun Park, Laboratory of Biochemistry and Metabolism, NIDDK, National Institutes of Health, Bethesda, MD 20892 **[145]**

Louis J. Picker, Department of Pathology, Stanford University, Stanford, CA 94305 **[91]**

George Poste, Smith Kline and French Laboratories, King of Prussia, PA 19406 **[173]**

Stanley B. Prusiner, Departments of Neurology, Biochemistry, and Biophysics, University of California, San Francisco, CA 94143-0518 **[35]**

James R. Rasmussen, Research and Development, Genzyme Corporation, Boston, MA 02111 **[159]**

Vernon N. Reinhold, Department of Nutrition, Division of Biological Sciences, Harvard School of Public Health, Boston, MA 02115 **[1]**

William L. Roberts, Department of Pharmacology, Case Western Reserve University, Cleveland, OH 44106 **[1]**

S. D. Rosen, Department of Anatomy, University of California, San Francisco, CA 94143-0452 **[69]**

Terrone L. Rosenberry, Department of Pharmacology, Case Western Reserve University, Cleveland, OH 44106 **[1]**

Sitthivet Santikarn, Department of Nutrition, Division of Biological Sciences, Harvard School of Public Health, Boston, MA 02115 **[1]**

C. Sapin, INSERM U178, 94807 Villejuif-Cedex, France **[127]**

Ashu Sharma, Department of Surgical Oncology, Roswell Park Memorial Institute, Buffalo, NY 14263 **[195]**

Barry D. Shur, Department of Biochemistry and Molecular Biology, The University of Texas M.D. Anderson Cancer Center, Houston, TX 77030 **[107]**

M. S. Singer, Department of Anatomy, University of California, San Francisco, CA 94143-0452 **[69]**

S. Stachel, Department of Cardiovascular Research, Genentech, Inc., San Francisco, CA 94080 **[69]**

Christopher Starr, Laboratory of Biochemistry and Metabolism, NIDDK, National Institutes of Health, Bethesda, MD 20892 **[145]**

L. M. Stoolman, Department of Pathology, University of Michigan, Ann Arbor, MI 48109 **[69]**

Philip R. Streeter, Department of Pathology, Stanford University, Stanford, CA 94305; present address: Systemix, Inc., Palo Alto, CA 94303 **[91]**

Daniel Sucato, Department of Surgical Oncology, Roswell Park Memorial Institute, Buffalo, NY 14263 **[195]**

Max D. Summers, Department of Entomology and Institute of Biosciences and Technology, Texas A & M University, College Station, Texas 77843 **[199]**

Jean-Pierre Toutant, Department of Pharmacology, Case Western Reserve University, Cleveland, OH 44106 [1]

D. D. True, Department of Anatomy, University of California, San Francisco, CA 94143-0452 [69]

Germain Trugnan, INSERM U178, 94807 Villejuif-Cedex, France [127]

S. J. Turco, Department of Biochemistry, University of Kentucky College of Medicine, Lexington, KY 40536 [23]

Mark L. Tykocinski, Institute of Pathology, Case Western Reserve University, Cleveland, OH 44106 [17]

Paul M. Wassarman, Department of Cell and Developmental Biology, Roche Institute of Molecular Biology, Roche Research Center, Nutley, NJ 07110 [59]

Joseph K. Welply, Department of Biological Sciences, Monsanto Company, St. Louis, MO 63198 [xiii]

Barbara Wolff, Laboratory of Biochemistry and Metabolism, NIDDK, National Institutes of Health, Bethesda, MD 20892 [145]

Barbara Woynarowska, Department of Experimental Therapeutics, Roswell Park Memorial Institute, Buffalo, NY 14263 [195]

T. A. Yednock, Department of Anatomy, University of California, San Francisco, CA 94143-0452; present address: Department of Microbiology and Immunology, University of California, San Francisco, CA 94143 [69]

Adel Youakim, Department of Biochemistry and Molecular Biology, The University of Texas M.D. Anderson Cancer Center, Houston, TX 77030 [107]

David F. H. Zhou, Department of Pathology, Stanford University, Stanford, CA 94305 [91]

A. Zweibaum, INSERM U178, 94807 Villejuif-Cedex, France [127]

Preface

The 1989 UCLA Symposium on **Glycobiology** was held in mid-January in Frisco, Colorado. The meeting marked the first UCLA Symposium dedicated to one of the most rapidly expanding fields in cell biology, a field that will certainly have a major impact on medical research and practice in the future. The conference was attended by over 180 scientists from across the globe and was organized into eight major sessions of talks with corresponding poster sessions. A pre-banquet address highlighted the week's events.

The meeting, and most of the articles published here, focused on the synthesis and structure of protein- and lipid-bound oligosaccharides and on the mechanisms by which they and carbohydrate-binding proteins contribute to biological recognition and communication. Several talks were devoted to the synthesis, structural characterization, and biological functions of glycophosphoinositol lipids that are covalently attached to the carboxyl terminus of a variety of proteins and anchor the protein to the outer leaflet of the plasma membrane. The pathway for synthesis of these lipids has been found to involve some of the precursors that are utilized for the synthesis of N-linked sugars. Additionally, a signal for the attachment of the lipid exists within the structure of the protein. Glycolipid anchors have a variety of biological functions; among these are to provide a mechanism for quickly shedding cell surface proteins, to protect cell surfaces from their environment and for cell–cell interactions. Parasite coat proteins are one protein type that contains glycolipid anchors. These and other glycolipid-anchored membrane proteins are susceptible to liberation from the membrane and release from the cell upon cleavage with phosphatidylinositol-specific phospholipases. In a process referred to as antigenic variation, parasites utilize this mechanism to alter dramatically their cell surface proteins in response to immunological challenges by the host. Another biological role of glycolipid anchors presented was that the glycan portion of the major lipophosphoglycan of the parasite Leishmania protects the parasite from degradative enzymes while the lipid portion inhibits protein kinase C. Additionally, a few cellular adhesion molecules are made in a form that contains a glycolipid anchor. The function of the glycolipid in these molecules is unknown.

Structural variations between the glycolipid anchors of the host proteins and those of parasites suggest that there may be parasite-specific glycosyltransferases that could be selectively inhibited and might have potential as antiparasiticides. The presence of two ethanolamine moieties within some of the glycolipid tails that were described suggested that more than one protein may be linked to another through a common glycolipid tail. Lastly, modified glycosylation, including perturbed glycolipid tails, may be associated with altered forms of proteins such as the cellular and scrapie prion protein that causes degenerative diseases such as Kuru, Creutzfeldt-Jakob disease, and Gertzmann-Sträussler syndrome.

A series of papers was presented on the role of lectin-like molecules and carbohydrate receptors in cell–cell recognition events. The characterization of an oligosaccharide on the surface of unfertilized mouse eggs that acts as the receptor for sperm was presented and shown to contain O-linked oligosaccharide with a terminal alpha-linked galactose. Cell adhesion molecules with lectin-like domains (Lec-Cams) have been cloned and evidence was presented that clearly indicated that they are involved in the homing of lymphocytes from the blood into secondary lymphoid organs. Cell surface forms of galactosyltransferase were shown to be a feature of many cell types and to be involved in cell recognition and cell migration along extracellular matrices.

Many additional papers were presented on the glycosylation of proteins and lipids in diseases such as rheumatoid arthritis or chorionic carcinoma. Changes in glycosylation accompanying normal tissue development, such as differentiation of enterocytes, as well as cell type and protein-specific glycosylation were covered. A session was devoted to concerns of pharmaceutical and biotechnology companies (i.e. drug targeting and recombinant protein expression) within this rapidly growing field. Finally, several posters presented work on the cloning of glycosyltransferases. The cloned enzymes should be valuable for testing the effects of temporal and organ-specific alterations in glycosylation within disease models of transgenic animals.

Many thanks are extended to Smith Kline & French Laboratories for its generous sponsorship of this meeting. Additional support was received from Monsanto Company; Glycomed, Inc.; Bio Carb, Inc.; and Genzyme Corporation. We also acknowledge the Symposia staff for their exceptional organizational support. We support and have received encouragement for a second symposium on glycobiology to be held in the near future.

Joe Welply
Ernie Jaworski

THE GLYCOINOSITOL PHOSPHOLIPID ANCHOR OF HUMAN ERYTHROCYTE ACETYLCHOLINESTERASE[1]

Terrone L. Rosenberry[2], William L. Roberts[2], Jean-Pierre Toutant[2], Sitthivet Santikarn[3], and Vernon N. Reinhold[3]

Department of Pharmacology[2], Case Western Reserve University, Cleveland, OH 44106 and Division of Biological Sciences[3], Harvard School of Public Health, Boston, MA 02115

ABSTRACT Several proteins, including acetylcholinesterases (AChEs) from mammalian erythrocytes, recently have been shown to be anchored on the extracellular face of plasma membranes exclusively by a covalently linked glycoinositol phospholipid. Chemical analyses have revealed that anchor components include an inositol phospholipid, glucosamine, mannose, and ethanolamine in amide linkage to the polypeptide C-terminus. Many of these anchored proteins are selectively released from cell membranes by purified bacterial phosphatidylinositol-specific phospholipase C (PIPLC). However, in some cells, particularly human erythrocytes, this release is only partial. For example, PIPLC releases only about 5% of the AChE from human erythrocytes. To investigate this PIPLC resistance, we cleaved the inositol phospholipid from the human erythrocyte AChE anchor with nitrous acid and examined its structure by fast atom bombardment mass spectrometry. This phospholipid contained 1-alkyl-2-acylglycerol in which the alkyl group was 18:0 and the predominant acyl group was 22:4 + 22:5. In addition, a novel palmitoyl group (16:0) was observed in direct acyl linkage to an inositol hydroxyl. This group was directly responsible

[1]This work was supported by Grants NS16577 and DK38181 from the National Institutes of Health, Grant RR01494 from the United States Public Health Service, Grant PCM8300342 from the National Science Foundation, and by grants from the Muscular Dystrophy Association.

for PIPLC resistance, because selective palmitate deacylation by ammonia methanolysis generated an inositol phospholipid that was cleaved by PIPLC to yield alkylacylglycerol and inositol-1-phosphate. Polyacrylamide gel electrophoresis in nondenaturing detergents resolves small amounts of AChE histochemical activity and permits rapid discrimination of detergent-binding AChE that retains anchor lipids from hydrophilic AChE that has lost anchor lipids. With this technique it was shown that deacylation of human erythrocyte AChE with 1M hydroxylamine at pH 10.7 yields a residual alkyllysoplasmanylinositol anchor that binds detergent micelles and is susceptible to PIPLC cleavage.

INTRODUCTION

The relationship of structural features of acetylcholinesterase (AChE) to the function of this enzyme in cholinergic transmission in the nervous system is not yet clear. Three distinct classes of membrane-bound AChEs have been identified, and very interesting differences in their membrane attachment structures have been found (1). One class of dimers labeled G_2 has been found to belong to a growing new list of integral membrane proteins that are anchored, not by a transmembrane peptide segment, but exclusively by a glycoinositol phospholipid linked covalently to the protein C-terminus. Although convincing evidence of glycoinositol phospholipid anchors has been obtained only in the last four years (2; see 3), some 30-40 proteins with such anchors now have been tentatively identified (4). All appear to reside on the extracellular face of the cell plasma membrane. G_2 AChEs with these anchors have been found in torpedo electric organ (5), insect heads (6-8) and mammalian erythrocytes (9-11).

Most of the proteins reported to have glycoinositol phospholipid anchors have been identified on the basis of their susceptibility to release from the cell surface by purified bacterial phosphatidylinositol-specific phospholipase C (PIPLC; 3). This enzyme is obtained from the culture medium of S. aureus or B. thuringiensis and appears free of protease contamination. No protein analyzed to date which is cleaved by PIPLC has failed to contain a glycoinositol phospholipid anchor. However, the converse is not true. In certain cell types, proteins with glycoinositol phospholipid anchors are not released or cleaved by PIPLC. In the following sections we outline the evidence that human erythrocyte AChE has such an anchor and show that the basis of the PIPLC resistance arises from a novel modification of the inositol phospholipid.

IDENTIFICATION OF A GLYCOINOSITOL PHOSPHOLIPID ANCHOR IN HUMAN ERYTHROCYTE ACETYLCHOLINESTERASE

Purification of Human Erythrocyte AChE and Identification of a Hydrophobic Domain.

AChE comprises only 0.01% of the protein in human erythrocyte membranes, and our strategy from the outset was to purify the enzyme in sufficient quantities to allow structural characterization of the membrane-binding domain. A tangential flow filtration procedure was introduced to remove hemolysate from membranes (12, 13) that in one day gives quantitative membrane recoveries with less than 1% hemoglobin contamination from 10 liters of outdated human erythrocytes. A rapid procedure was developed for extraction of the enzyme with Triton X-100 and batch adsorption to acridinium affinity resin, and elution from the washed enzyme-resin complex gives AChE free of protein contaminants in overall yields of about 50% (13). The purified enzyme is a disulfide-linked dimer of apparently identical catalytic subunits, and quantitative measurements of [^{3}H]Triton X-100 binding indicated that each subunit interacts with one detergent micelle (13). This AChE is classified as an amphiphilic protein because digestion with papain separated active hydrophilic AChE from its detergent-binding domain (14). The intact amphiphilic AChE could be reconstituted into phospholipid liposome membranes by detergent dialysis procedures, and more than 80% of the liposomal AChE was released by papain as the active hydrophilic AChE fragment (15). Polyacrylamide gel electrophoresis in sodium dodecyl sulfate (SDS-PAGE) indicated that the intact subunit mass of about 70 kDa was decreased only by an apparent 2 kDa when the hydrophilic AChE fragment was generated by papain (14, 15). Because of the similar size of intact enzyme and the hydrophilic fragment, it was concluded that papain had released a small hydrophobic fragment from the subunit N- or C-terminus that contained the membrane-anchoring domain.

Isolation of the AChE Membrane Anchor and Determination of Anchor Components.

To pursue the structure of this membrane anchor, specific radioisotopic labels were needed that would permit isolation of the hydrophobic fragment produced by papain. Two useful radiolabeling procedures that were developed are illustrated in Figure 1. Reductive methylation with

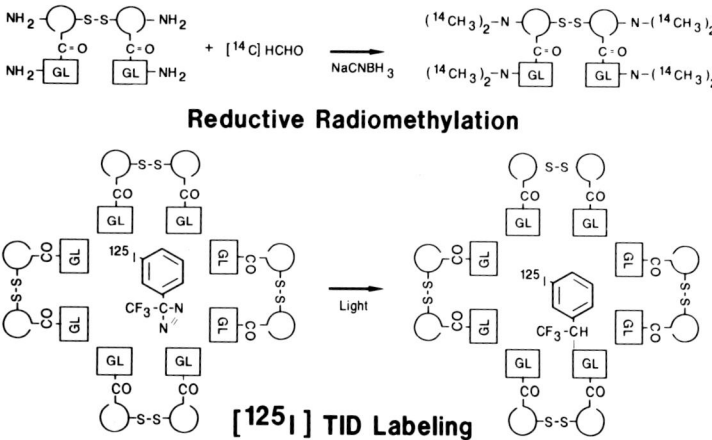

FIGURE 1. Radiolabeling reagents for purified AChE. The dimeric G_2 AChE structure is represented by circles joined by a disulfide bond (S-S) and linked to the glycoinositol phospholipid (GL) at the protein C-terminus (C=O).

[^{14}C]HCHO and NaCNBH$_3$ generates radiolabeled mono- and dimethylated amines that are stable to acid hydrolysis and detectable on an amino acid analyzer (16). Radiomethylated amine groups identified in AChE are discussed below. The reagent 3-(trifluoromethyl)-3-(m-[^{125}I]iodophenyl)diazirine ([^{125}I]TID) partitions selectively into the hydrophobic phase of reaction mixtures and, on photoactivation, reacts covalently with molecules in this phase (17). When Triton X-100 was removed from purified AChE, interaction of the enzyme hydrophobic domains induced the formation of miceller aggregates illustrated in Figure 1. [^{125}I]TID partitioned effectively into these aggregates where it labeled the enzyme, and the labeled enzyme was repurified by affinity chromatography to remove residual labeled contaminants (18;

see 14). The [^{125}I]TID was selective for the anchor, as papain digestion of the radiolabeled AChE quantitatively released the label in the small hydrophobic fragment. This fragment was purified in a three-step procedure. Digestion was conducted with papain attached to Sepharose CL-4B in Triton X-100, and the supernatant was adsorbed to acridinium affinity resin to remove the hydrophilic enzyme fragment. The nonretained ^{125}I-labeled fragment in Triton X-100 micelles was then chromatographed on Sepharose CL-6B, and detergent was removed by further chromatography on Sephadex LH-60 in an ethanol-formic acid solvent (18).

It was initially anticipated that the AChE anchor would correspond to a short polypeptide segment of hydrophobic amino acids similar to those reported for other amphiphilic proteins like cytochrome b_5 (19). However, chemical analyses of the isolated [^{125}I]TID-labeled fragment produced by papain indicated a more novel anchor structure. The only amino acids detected were 1 mole of histidine and 1 mole of glycine per mole of fragment (18), and gas-liquid chromatography (GLC) and GLC-mass spectrometry (GLC-MS) following methanolysis revealed that the fragment contained 2 residues of fatty acids (9). Palmitate (16:0) was the primary saturated fatty acid, and the predominant unsaturated fatty acids were the relatively unusual 22:4 and 22:5. GLC analysis following hydrolysis in 6N HCl detected about 1 residue of inositol in this fragment (20). Radiomethylation contributed three more important pieces of information about this AChE anchor. First, it revealed 1-2 ethanolamine and 1 glucosamine residues with free primary amino groups in the small fragment produced by papain (21). Second, the anchor was localized to the subunit C-terminus. Radiomethylated N-terminal Glx in the intact AChE was quantitatively retained in the hydrophilic AChE fragment generated by papain, and manual Edman sequencing in combination with radiomethylation revealed the sequence of the small fragment to be His-Gly-ethanolamine-glycolipid (21). Third, this ethanolamine in amide linkage to the C-terminal Gly was thus identified as the covalent bridge to the remainder of the glycolipid anchor.

STRUCTURE OF THE HUMAN ERYTHROCYTE ACETYLCHOLINESTERASE ANCHOR AND DEMONSTRATION THAT PALMITOYLATION OF INOSITOL CONVEYS RESISTANCE TO PIPLC

By 1985, the above work from our laboratory could be compared 1) with observations from the Silman laboratory that G_2 AChE was released from torpedo electric organ membranes by PIPLC and was covalently linked to inositol (5, 22) and 2) with a report from the Cross laboratory that trypanosome variant surface glycoproteins (VSGs) were attached to the surface membrane by covalently linked dimyristoylphosphatidylinositol (2). A perplexing observation, however, was that PIPLC did not release AChE from human erythrocytes even though it did release AChE from erythrocytes of other mammalian species including the cow (23). As a first step in resolving this problem, AChEs were purified from both bovine and human erythrocytes, labeled with [^{125}I]TID, and repurified. As shown in Figure 2, the

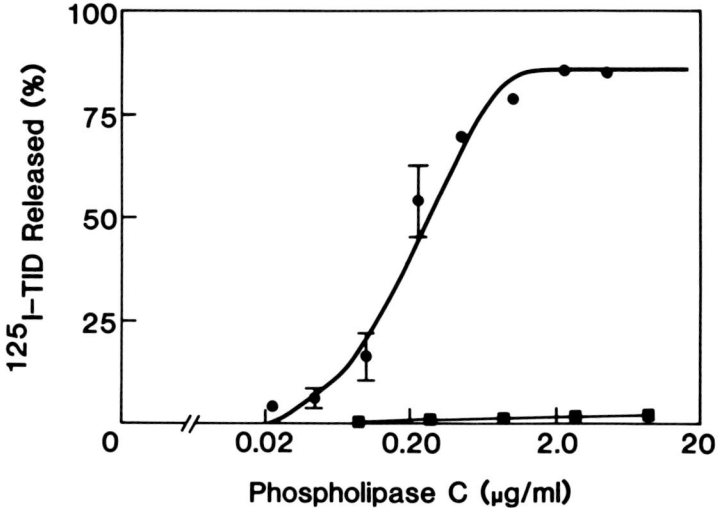

FIGURE 2. Treatment of AChEs with PIPLC. [^{125}I]TID-labeled AChEs from bovine (circles) and human erythrocytes (squares) (3000 dpm per sample) were incubated with purified S. aureus PIPLC in 0.1% sodium deoxycholate, 20 mM sodium phosphate (pH 7) for 90 min at 37° C. Anchor cleavage corresponds to percent release of ^{125}I label from AChE protein bands to the dye front following SDS-PAGE (20).

labeled AChEs were then incubated under identical conditions with varying concentrations of PIPLC and cleavage was compared. PIPLC at greater than 2 ug/ml released about 85% of the ^{125}I label from the bovine AChE but less than 5% from the human AChE. Thus the resistance of the human AChE anchor to PIPLC was related to the anchor structure itself and did not involve other features of human erythrocyte membranes. To obtain more information about these AChE anchor structures, lipid-containing fragments were generated by several procedures and identified by thin-layer chromatography (TLC). These procedures and the deduced inositol phospholipid structure are illustrated in Figure 3. The

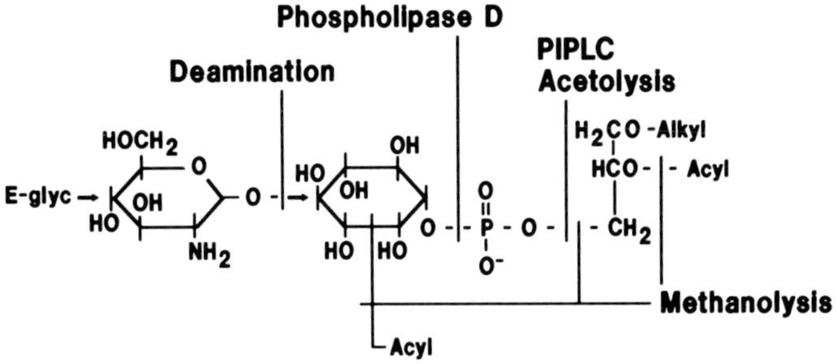

FIGURE 3. Procedures that generate fragments useful for analysis of the AChE glycoinositol phospholipid anchors.

analysis was very sensitive because lipid fragments from [^{125}I]TID-labeled phosphatidylinositol as well as from the labeled AChEs showed TLC mobilities similar to unlabeled lipid standards (20). For example, labeled fatty acids released by base hydrolysis and fatty acid methyl esters produced by acidic methanolysis from both AChEs were identified by their correspondence to unlabeled lipid standards on TLC (see Figure 4A). In contrast to the dimyristoylglycerol released from trypanosome VSG anchors by PIPLC (2), the labeled lipid released by PIPLC from the [^{125}I]TID-labeled bovine AChE in Figure 2 migrated slightly faster than standard diacylglycerols on TLC. This discrepency led to a collaboration with the Kuksis laboratory and to a determination by GLC and high performance liquid chromatography (HPLC) that 96% of the released lipid was 1-alkyl-2-acylglycerol with predominantly 18:0 alkyl and 18:0 acyl groups

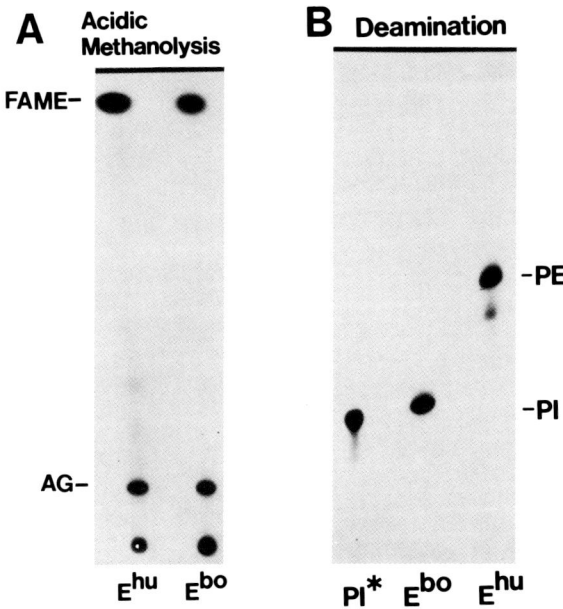

FIGURE 4. Silica TLC analysis of lipid fragments from [^{125}I]TID-labeled bovine (E^{bo}) and human (E^{hu}) erythrocyte AChEs. A. Labeled AChEs were subjected to acidic methanolysis, extracted with chloroform, and chromatographed in hexane/2-propanol (96:4) (10). Unlabeled standards were 1-hexadecylglycerol (AG) and methyl palmitate (FAME). B. Labeled AChEs were treated with nitrous acid, extracted with chloroform/methanol (2:1), and chromatographed in chloroform/methanol/water (65:25:4) (20). Unlabeled standards were phosphatidylethanolamine (PE) and phosphatidylinositol (PI). PI* represents [^{125}I]TID-labeled PI spotted directly on the plate without nitrous acid treatment. Horizontal lines mark the solvent fronts. Labeled lipids were detected by autoradiography of the TLC plate. The lowest spots in panel A indicate the origin and may result from the degradation of [^{125}I]TID.

(24). Acidic methanolysis of the [^{125}I]TID-labeled bovine and human AChEs released labeled alkylglycerols as well as fatty acid methyl esters from both enzymes (Figure 4A), suggesting alkylacylglyerol components in both enzyme anchors.

To clarify the structure of the inositol phospholipid in human erythrocyte AChE, more extensive chemical analyses were conducted. Quantitative GLC analysis after acidic methanolysis and acetylation confirmed the 2 residues of fatty acids determined previously as well as 0.8 residues of 18:0 and 18:1 alkylglycerols per anchor. This result was surprising because a diradylglycerol can account for only two fatty alkyl or acyl groups. To confirm the presence of a diradylglycerol, acetolysis was employed to cleave the putative phosphate ester bond in the human AChE that appears resistant to PIPLC (Figure 3). While this reaction led to some isomerization and side-product formation, the Kuksis laboratory determined that the released lipid acetolysis products were derived exclusively from 1-alkyl-2-acylglycerols, 83% of which contained an 18:0 or 18:1 alkyl group and a 22:4, 22:5 or 22:6 acyl group (10). A clue to the location of the second fatty acyl group was provided by nitrous acid deamination, a particularly useful cleavage noted in Figure 3 that breaks the C-1 glycosidic bond of the glucosamine residue and releases inositol phospholipid. While deamination of [^{125}I]TID-labeled bovine AChE released a labeled lipid with nearly the same TLC mobility as the labeled phosphatidylinositol control, the deamination product from the labeled human AChE was an unusually nonpolar inositol phospholipid with much greater TLC mobility (Figure 4B; 20). Thus it appeared that the second fatty acyl group was attached at some site on this inositol phospholipid. This location was confirmed by fast atom bombardment mass spectrometry (FAB-MS). The unusual inositol phospholipid released by deamination was purified by HPLC and shown to be a plasmanylinositol (an analog of phosphatidylinositol in which glycerol has one O-linked alkyl and one O-linked acyl group) with a second residue of fatty acid, primarily palmitate, in direct acyl linkage to an inositol hydroxyl group (11). Furthermore, the intact human erythrocyte AChE anchor was produced by proteolysis of this AChE with Pronase, deacylated by base hydrolysis, purified by HPLC and analyzed by FAB-MS (11). The analyses indicated the E^{hu} AChE anchor structure shown in Figure 5, where it is compared to that reported for the VSG anchor (25). Both anchors contain a similar backbone of 3 linear hexose residues with a phosphodiester linkage at their nonreducing terminus to an ethanolamine in amide linkage to the C-terminal amino acid and a glucosamine at their reducing terminus linked to the inositol phospholipid. This phospholipid in VSG is simply dimyristoylphosphatidylinositol, while in human AChE it

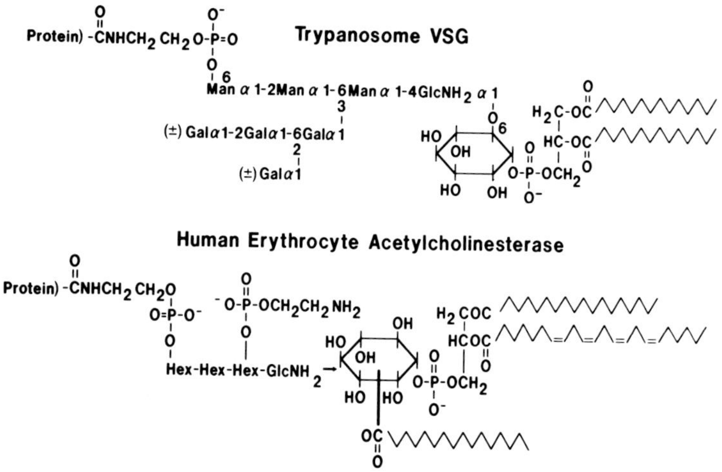

Trypanosome VSG

Protein)-CNHCH$_2$CH$_2$O-P(=O)(O$^-$)-O-[Manα1-2Manα1-6Manα1-4GlcNH$_2\alpha$1 with branches at position 3: (±)Galα1-2Galα1-6Galα1, position 2: (±)Galα1]-inositol-O-P(=O)(O$^-$)-OCH$_2$-CH(O-CO-acyl)-CH$_2$-O-CO-acyl

Human Erythrocyte Acetylcholinesterase

Protein)-CNHCH$_2$CH$_2$O-P(=O)(O$^-$)-O-P(=O)(O$^-$)-OCH$_2$CH$_2$NH$_2$ - Hex-Hex-Hex-GlcNH$_2$ → inositol(OC(=O)-alkyl)-O-P(=O)(O$^-$)-OCH$_2$-CH(O-CO-alkenyl)-CH$_2$-O-alkyl

FIGURE 5. Schematic representations of anchor structures. ∽ represents the carbon atom backbone of the alkyl and acyl chains. Upper panel: membrane anchor of *T. brucei* MITat 1.4 VSG (from 2, 25). Lower panel: in the membrane anchor of human erythrocyte AChE, the positions of glycosidic and hexose phosphate linkages and the anomeric configurations remain to be determined (from 11).

includes the alkylacylglycerol and palmitoylated inositol noted above. An additional phosphorylethanolamine is attached to the hexose adjacent to the glucosamine in the AChE anchor, while in the VSG anchor a branching galactose side chain is found at this location.

Our observation of palmitoylation of inositol in the human erythrocyte AChE anchor to our knowledge is the first documented case of endogenous fatty acid acylation of inositol. Furthermore, this palmitoyl group was shown to be responsible for the resistance of the AChE inositol phospholipid to PIPLC, because treatment of the deamination product with NH_3-saturated methanol removed the palmitoyl group and permitted release of alkylacylglycerol and inositol-1-phosphate on further treatment with PIPLC. In contrast, an anchor-specific phospholipase D observed in mammalian sera (26, 27) cleaved the intact anchors of both the human and the bovine AChEs at the inositol-phosphate bond (Figure 3) with release of plasmanic acid (10). This cleavage was observed only in the presence of nonionic detergents.

A TECHNIQUE FOR DETECTING ACHE ANCHOR RESISTANCE TO PIPLC THAT ARISES FROM ANCHOR ACYLATION

In all glycoinositol phospholipid-anchored proteins which have been studied, the inositol phospholipid is the only nonionic detergent-binding domain. Cleavage of the anchor by PIPLC removes the hydrophobic diacyl- or alkyl-acylglycerol and generates a hydrophilic protein which retains the glycan portion of the anchor. One technique that can demonstrate this loss of detergent-binding capacity is polyacrylamide gel electrophoresis in nonionic detergents (28). Since associated detergent micelles contribute substantially to the Stokes radii of intact anchored proteins, loss of detergent binding can generate a several-fold increase in the electrophoretic migration of the cleaved proteins. In contrast, anchor cleavage often is undetectable by SDS-PAGE. The enzyme activity of AChE is retained in nondenaturing gels, and AChE migration can be assessed with great sensitivity by a histochemical stain (29) even in crude cell extracts. Cleavage of bovine erythrocyte AChE by PIPLC is demonstrated in Figure 6A (lane b),

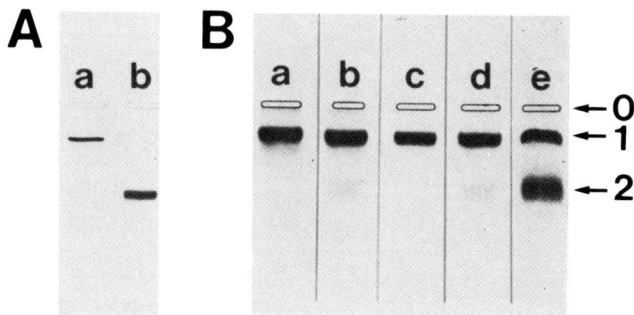

FIGURE 6. Histochemical detection of AChE following electrophoresis in 7.5% polyacrylamide gels with 0.5% Triton X-100 at pH 8.9 (30). A. Bovine AChE. B. Human AChE. Lanes a: control samples. Lanes b: samples incubated with PIPLC (5 ug/ml) for 1 hr at 37° C. Lane c: sample adjusted to pH 10.7 in 0.8 M hydroxylamine for 2 hr at 25° C. Lane d: PIPLC incubation followed by hydroxylamine treatment. Lane e: hydroxylamine treatment followed by dialysis at pH 7 for 1 hr and then by PIPLC incubation. Arrow 0, loading well; arrow 1, intact AChE; arrow 2, cleaved AChE. Above pH 11, AChE activity is lost irreversibly.

where the cleaved AChE migrates more than twice as far as the control intact enzyme in Figure 6A (lane a). Moreover, this technique has been extended to provide the following sensitive assay for the presence of lipid components in the human erythrocyte AChE anchor (30). Palmitoylation of inositol prevents PIPLC cleavage of this anchor, as described in the preceding section and demonstrated for the intact AChE in Figure 6B, lane b. However, this fatty acid as well as that at the 2-position of glycerol can be partially deacylated by hydroxylamine at pH 10.7. Partial removal of these fatty acids has no effect on the electrophoretic mobility of human AChE (Figure 6B, lane c), an indication that the alkylglycerol group alone is sufficient to retain the binding of one Triton X-100 micelle per AChE subunit. However, PIPLC digestion following hydroxylamine treatment now can release this alkylglycerol group and generate a substantial amount of hydrophilic AChE (Figure 6B, lane e). Hydroxylamine cleaves ester linkages quite selectively at high pH. While this technique cannot prove that an anchored protein is resistant to PIPLC by fatty acid acylation specifically on inositol, it can provide compelling evidence that resistance arises from acylation at some anchor site.

SUSCEPTIBILITY TO PHOSPHOLIPASE CLEAVAGE MAY REGULATE THE FUNCTIONS OF GLYCOINOSITOL PHOSPHOLIPID ANCHORS

Aside from its obvious role in attaching proteins to cell membranes, biological functions that necessitate this glycolipid anchoring mechanism instead of a transmembrane peptide are unclear. Endogenous phospholipases C in trypanosomes (31) and rat liver (32) and D in mammalian serum (26, 27) have been discovered that are much more specific for anchor glycoinositol phospholipids than other inositol phospholipids, and the existence of these enzymes has suggested a functional role for endogenous phospholipase cleavage. This suggestion is supported 1) by one example of an anchored latent protease in malarial parasites that is activated by phospholipase C cleavage (33) and 2) by evidence that insulin can activate endogenous phospholipases that cleave both protein anchors and free glycoinositol phospholipids with similar structures. The glycans released from the free glycolipids appear to act as insulin second messengers. They regulate the activities of insulin-sensitive enzymes like cAMP phosphodiesterase, adenylate cyclase, pyruvate dehydrogenase and phospholipid methyltransferase in cell-free assays, and they show insulin-mimetic effects on glucose utilization and lipolysis in intact adipocytes (34). Recognition that palmitoylation of human erythrocyte AChE conveys resistance to PIPLC further suggests that regulation of susceptibility to phospholipase cleavage may be important in the control of glycoinositol phospholipid function.

The susceptibility of a glycoinositol phospholipid-anchored protein to phospholipase C cleavage appears to vary in a cell-specific manner. Two other anchored proteins, DAF (35, 36) and LFA-3 (37), also are resistant to PIPLC cleavage in human erythrocytes but susceptible to PIPLC in other blood cells, and preliminary data indicate that the erythrocyte DAF anchor also is palmitoylated on inositol (38). These observations are consistent with the notion that all anchored proteins in a given cell type have similar glycoinositol phospholipid structures. Studies of the biosynthesis of glycoinositol phospholipid-anchored proteins indicate how this similarity could arise. cDNA sequences for proteins linked to these anchors predict C-terminal amino acid sequences that extend 17-31 residues beyond those observed in the mature protein (4). The extension appears to function as a C-terminal "signal sequence" that is cleaved when essentially complete anchors that include the phospholipid are attached in the rough endoplasmic reticulum very shortly after translation (39, 40). Rapid anchor attachment suggests that anchor precursor glycoinositol phospholipids are assembled en bloc prior to their transfer to the protein C-terminus, and evidence to support such assembly has been obtained in the trypanosome system. A putative glycolipid precursor to the trypanosome VSG anchor was labeled biosynthetically with [^{3}H]myristate and identified by characteristic reagent cleavages noted in Figure 3 (41). Of further interest, a second [^{3}H]myristate-labeled glycoinositol phospholipid also was detected in trypanosome extracts (41) which is insensitive to PIPLC because of an apparent modification by an acyl group (42)[1]. This similarity to the palmitoylation of the human erythrocyte AChE anchor raises the interesting possibility that both PIPLC-sensitive and fatty acylated, PIPLC-resistant anchor precursors may coexist in mammalian cells. The choice of which precursor is attached then would be regulated according to the cell type by a mechanism yet to be determined.

[1]Personal communication, GAM Cross, AK Menon, S Mayor, RT Schwarz.

REFERENCES

1. Inestrosa NC, Roberts WL, Marshall TL, Rosenberry TL (1987). J Biol Chem 262:4441-4444.
2. Ferguson MAJ, Low MG, Cross GAM (1985). J Biol Chem 260:14547-14555.
3. Low MG (1987). Biochem J 244:1-13.
4. Ferguson MAJ, Williams AF (1988). Annu Rev Biochem 57:285-320.
5. Futerman AH, Fiorini RM, Roth E, Low MG, Silman I (1985). Biochem J 226:369-377.
6. Gnagey AL, Forte M, Rosenberry TL (1987). J Biol Chem 262:13290-13298.
7. Fournier D, Berge J-B, Cardoso de Almeida M-L, Bordier C (1988). J Neurochem 50:1158-1163.
8. Haas R, Marshall TL, Rosenberry TL (1988). Biochemistry 27:6453-6457.
9. Roberts WL, Rosenberry TL (1985). Biochem Biophys Res Commun 133:621-627.
10. Roberts WL, Myher JJ, Kuksis A, Low MG, Rosenberry TL (1988). J Biol Chem 263:18766-18775.
11. Roberts WL, Santikarn S, Reinhold VN, Rosenberry TL (1988). J Biol Chem 263:18776-18784.
12. Rosenberry TL, Chen J, Lee M, Onigman P (1981). J Biochem Biophys Meths 4:39-48.
13. Rosenberry TL, Scoggin DM (1984). J Biol Chem 259:5643-5652.
14. Dutta-Choudhury TA, Rosenberry TL (1984). J Biol Chem 259:5653-5660.
15. Kim BH, Rosenberry TL (1985). Biochemistry 24: 3586-3592.
16. Haas R, Rosenberry TL (1985). Analyt Biochem 148:154-162.
17. Brunner J, Semenza G (1981). Biochemistry 20: 7174-7182.
18. Roberts WL, Rosenberry TL (1986). Biochemistry 25:3091-3098.
19. Fleming PJ, Dailey HA, Corcoran D, Strittmatter P (1978). J Biol Chem 253:5369-5372.
20. Roberts WL, Kim BH, Rosenberry TL (1987). Proc Natl Acad Sci USA 84:7817-7821.
21. Haas R, Brandt PT, Knight J, Rosenberry TL (1986). Biochemistry 25:3098-3105.
22. Futerman AH, Low MG, Ackermann KE, Sherman WR, Silman I (1985). Biochem Biophys Res Commun 129:312-317.
23. Low MG, Finean JB (1977). FEBS Lett 82:143-146.

24. Roberts WL, Myher JJ, Kuksis A, Rosenberry TL (1988). Biochem Biophys Res Commun 150:271-277.
25. Ferguson MAJ, Homans SW, Dwek RA,, Rademacher TW (1988). Science 239:753-759.
26. Davitz MA, Hereld D, Shak S, Krakow J, Englund PT, Nussenzweig V (1987). Science 238:81-84.
27. Low MG, Prasad ARS (1988). Proc Natl Acad Sci USA 85:980-984.
28. Arpagaus M, Toutant J-P (1985). Neurochem Int 7:793-804.
29. Karnovsky MJ, Roots L (1964). J Histochem Cytochem 12:219-222.
30. Toutant J-P, Roberts WL, Murray NR, Rosenberry TL (1989). Eur J Biochem, in press.
31. Bulow R, Overath P (1987). J Biol Chem 261:11918-11923.
32. Fox JA, Soliz NM, Saltiel AR (1987). Proc Natl Acad Sci USA 84:2663-2667.
33. Braun-Breton C, Rosenberry TL, Pereira da Silva L (1988). Nature 332:457-459.
34. Low MG, Saltiel AR (1988). Science 239:268-275.
35. Davitz MA, Low MG, Nussenzweig V (1986). J Exp Med 163:1150-1161.
36. Medof ME, Walter EI, Roberts WL, Haas R, Rosenberry TL (1986). Biochemistry 25:6740-6747.
37. Selvaraj P, Dustin ML, Silber R, Low MG, Springer TA (1987). J Exp Med 166:1011-1025.
38. Walter EI, Roberts WL, Rosenberry TL, Medof ME (1987). Fed Proc 46:772.
39. Bangs JD, Hereld D, Krakow JL, Hart GW, Englund PT (1985). Proc Natl Acad Sci USA 82: 3207-3211.
40. Conzelmann A, Spiazzi A, Bron C (1987). Biochem J 246:605-610.
41. Krakow JL, Hereld D, Bangs JD, Hart GW, Englund PT (1986). J Biol Chem 261:12147-12153.
42. Masterson WJ, Doering TL, Hart GW, Englund PT (1989). Cell, in press.

THE CYTOPLASMIC EXTENSION AS A DETERMINANT FOR GLYCOINOSITOLPHOSPHOLIPID ANCHOR SUBSTITUTION[1]

M. Edward Medof and Mark L. Tykocinski

Institute of Pathology, Case Western Reserve University, Cleveland, Ohio 44106

ABSTRACT Oligonucleotide cassette mutagenesis of sequences encoding lymphocyte CD8 and the complement decay-accelerating factor (DAF) was employed to identify the signal(s) that direct glycoinositolphospholipid (GPL) processing for selected surface proteins. Removal of 3' sequence from CD8 and transfer of this sequence to DAF showed that the absence of cytoplasmic charged residues is an obligatory, but not sufficient, requirement for directing GPL-anchor processing.

INTRODUCTION

A number of eukaryotic cell surface proteins are membrane anchored by covalently-linked glycoinositolphospholipid (GPL) moieties (reviewed in refs. 1 and 2). The mRNA transcripts which encode these proteins contain sequences for C-terminal hydrophobic peptides. Biosynthetic analyses have shown that immediately following emergence of nascent protein from ribosomes (3, 4), these hydrophobic peptides are excised and the GPL-anchoring moieties are substituted en bloc.

Previous studies have shown that the information which directs GPL-anchor processing resides in mRNA 3' end sequence (5, 6). The signals that focus this GPL processing only upon particular proteins and not others, however, are unknown. There are no consensus sequences apparent in mRNA 3' end sequence of different GPL-proteins. A particular primary sequence may not be critical, however, in that recent studies

---

[1] This work was supported by NIH grants AI23598, P01DK38181, and CA47566.

have shown that randomly generated hydrophobic sequences can substitute (7).

A feature that is shared by mRNAs encoding most GPL-anchored proteins is the absence of mRNA sequence for a cytoplasmic extension. One exception to this rule is murine Qa-2 which contains three cytoplasmic charged residues, but a charged amino acid (aspartic acid) is found in the transmembrane segment (8). Anchorage instability of nascent protein in the rough ER arising from the absence of cytoplasmic charged sequence (generally including basic amino acids at the cytoplasmic face of the cell membrane) could designate these proteins for GPL modification. To test this idea, the effects on the GPL-anchoring of the complement decay-accelerating factor (DAF) of appending cytoplasmic sequence and on conventional anchoring of lymphocyte CD8 of deleting cytoplasmic sequence were examined.

MATERIALS AND METHODS

Plasmid Constructions.

Preparation of CD8 and CD8·DAF cDNAs (corresponding to the native CD8 and to a chimeric protein encompassing the complete extracellular region of CD8 and the 3' end sequence of DAF) and their insertion into REP2 [an Epstein-Barr virus (EBV)-based episomal replicon in which the Rous sarcoma virus 3' long terminal repeat directs transcription] have been described (5).

CD8·DAF·CD8 was prepared by annealing the two synthetic oligonucleotides, 5' CATGGGGTTGCTGACTAGGAACCGAAGACGCGTCTAGA 3' and 5' CATGTCTAGACGCGTCTTCGGTTCCTAGTCAGCAAGCC 3' to each other and ligating the resulting duplex (containing Nco-I compatible overhanging ends) into a unique Nco-I site near the 3' end of the coding sequence of CD8·DAF in Bluescript (BT) (5); CD8·cyto- was assembled in an analogous fashion employing the two synthetic oligonucleotides, 5' ATCTACATCT GGGCGCCCTTGGCCGGGACTTGTGGGGTCCTTCTCCTGTCA 3' and 5' CGATTAC GCGTGGTTGCAGTAAAGGGTGATACCCAGTGACAGGAGAAGGAC 3'. These oligonucleotides were annealed, overhangs were filled in with Klenow fragment, and the resulting duplex was blunt-end ligated into the unique EcoRV site at the junction of the extracellular and transmembrane domains of CD8 in BT (5). The CD8·DAF·CD8 and CD8·cyto- cartridges were mobilized with KpnI and BamHl and inserted into corresponding sites of REP2. Oligonucleotide duplexes for both constructions incorporated unique Mlu1 reporter sites near their downstream ends for identification and orientation purposes. To confirm that the coding sequences were in-frame at insertion sites, junctions of each were sequenced by the dideoxy chain termination method (9).

Transfection Procedures.

K562 cells were electroporated using a Promega Biotec X-Cell 450 electroporator (Madison, WI) (5). Stable transfectants were selected with 0.2 mg/ml Hygromycin B (Calbiochem Inc.) 72 hrs post-transfection.

Flow Cytometric Analyses.

Transfectants (10^6 cells) were incubated for 30 min at 37°C with B. thuringenesis phosphatidylinositol-specific phospholipase C (PI-PLC) (M. Low, Columbia Univ., NY) diluted 1:50 in PBS or with PBS alone. Washed cells were then stained with 5 µg/ml anti-CD8 or nonrelevant MOPC mAB followed by a 1:50 dilution of FITC-labeled goat anti-Ig (Cappel Products, CA). The stained cells were analyzed on an Ortho Diagnostics Cytofluorograph as previously described (5).

RESULTS AND DISCUSSION

Previous studies established that chimeric CD8 protein encoded by CD8·DAF in REP2 undergoes GPL processing in (strain 253) K562 cells and that the surface protein is cleaved >80% by PI-PLC digestion (5). To determine whether the absence in nascent protein of charged amino acid residues characteristic of cytoplasmic extensions is an essential signal for GPL-anchor processing, the effect on anchoring of addition of a short stretch of basic amino acids to the end of DAF's hydrophobic peptide was assessed. For this purpose 6 amino acids (which are present at the start of the cytoplasmic extension of native CD8), 5 of which are basic, were appended to the 17 hydrophobic amino acid residues that occur naturally in the C-terminal DAF extension peptide. The resulting construct CD8·DAF·CD8 (Fig. 1) in REP2 was introduced into (strain 253) K562 cells and stable hygromycin-resistant transfectants selected.

Multiple cell transfectants expressing the CD8·DAF·CD8 gene product as determined by flow cytometry with anti-CD8 mABs (see Methods) were identified. The PI-PLC sensitivity of the CD8 antigen on 3 of these transfectants was compared

CD8

    170   190   210
PLAGTCGVLLLSLVITLYCNH | RNRRR VCKCPRPVVKSGDKPSLSARYV

CD8·cyto-

    170
PLAGTCGVLLLSLVITLYCNH | A

CD8·DAF·CD8

    340
GHTCFTLTGLLGTLVTMGLLT | RNRRRV

FIGURE 1. C-terminal amino acid sequences of CD8·cyto- and CD8·DAF·CD8, and their relationship to the C-terminal sequence of CD8. A horizontal bar is drawn over the six amino acids in CD8 that have been appended to the terminal amino acid of DAF in CD8·DAF·CD8. The amino acid positions within CD8 (for CD8 and CD8·cyto-) and DAF (for CD8·DAF·CD8) are shown.

to that of CD8 proteins on control K562 transfectants bearing CD8/REP2 (conventionally anchored CD8) and bearing CD8·DAF/REP2 (GPL-anchored CD8). The results are summarized in Table 1.

As found previously (5), CD8 antigen on the CD8/REP2 and CD8·DAF/REP2-containing K562 controls exhibited PI-PLC resistance and sensitivity, respectively. In contrast to the latter control cell, CD8 protein on all three CD8·DAF·CD8/REP2 transfectants was PI-PLC resistant.

To establish whether absence of a cytoplasmic extension is the only required signal for GPL-anchor processing, the effect on anchoring of creating a truncated CD8 protein in which the cytoplasmic extension on native CD8 is deleted was assessed. For this purpose a stop codon was introduced just downstream of the 26 amino acid long hydrophobic transmembrane sequence of native CD8. The resulting construct (CD8·cyto-, Fig. 1) was introduced into the same cell line and stable hygromycin-resistant CD8·cyto-/REP2 K562 transfectants were isolated as above. In contrast to the CD8·DAF·CD8/REP2 transfectants, however, all CD8·cyto-/REP2 transfectants failed to exhibit surface CD8 expression. Northern analyses of mRNA from the cells hybridized with [^{32}P]-labeled CD8 cDNA, however, revealed the presence of CD8·cyto- mRNA transcripts (not shown).

TABLE 1
PI-PLC SENSITIVITY OF K562 TRANSFECTANTS

		Mean CD8 fluorescence intensity[a]		Percent release
		Buffer	PI-PLC	
Controls				
CD8/REP2	#1	141	128	9
CD8·DAF/REP2	#1	400	29	93
	#2	340	17	95
Samples				
CD8·DAF·CD8/REP2	#13	312	298	4
	#16	627	549	12

[a]Transfectants (10^6 cells) were incubated at 37°C for 60 min with (B. Thuringenesis) PI-PLC (1:100) in 100 μl PBS or with an equal volume of PBS buffer alone. After centrifugation, cells were stained with anti-CD8 murine monoclonal antibody followed by FITC-labeled goat anti-murine Ig and analyzed by flow cytometry.

The demonstration that CD8 protein on CD8·DAF·CD8 transfectants was uniformly PI-PLC resistant indicates that a nonhydrophobic cytoplasmic extension comprised of as few as 6 amino acid residues is sufficient to prevent GPL-anchor substitution and confer conventional anchoring. The finding, however, that CD8 protein on CD8·cyto- transfectants uniformly failed to process normally and anchor on the surface membrane indicates that the absence of a cytoplasmic extension is not in and of itself a sufficient signal for GPL-anchor substitution. This could be a consequence of the presence in the CD8 transmembrane domain of features which interfere with directing the protein along the GPL processing pathway or alternatively could reflect the absence in upstream CD8 sequences of a suitable acceptor site for linkage of the GPL moiety.
Taken together our results suggest that the absence of a cytoplasmic extension constitutes a part, but not all, of the signal for GPL substitution. Site-specific mutagenesis strategies should provide additional insights into the primary sequence requirements for this process. Ultimately,

however, a precise resolution of this issue may have to await the molecular characterization of the enzymes responsible for GPL-anchor substitution and insights into the three-dimensional structure of the enzymes and their heterogeneous protein substrates.

ACKNOWLEDGEMENTS

We thank Dr. Christopher Fauer for his help in the gene transfections, Anne Litrizza and Robert R. Getty for their technical assistance, and Beth Finke and Sara Cechner for help in manuscript preparation.

REFERENCES

1. Cross GAM (1987). Eukaryotic protein modification and membrane attachment via phosphatidylinositol. Cell 48:179.
2. Low MG (1989). Glycosyl-phosphatidylinositol: a versatile anchor for cell surface proteins. FASEB J 3:1600.
3. Medof ME, Walter EI, Roberts WL, Haas R, Rosenberry TL (1986). Decay accelerating factor of complement is anchored to cells by a C-terminal glycolipid. Biochemistry 25:6740.
4. Bangs JD, Andrews NW, Hart GW, Englund PT (1986). Post-translational modification and intracellular transport of a trypanosome variant surface glycoprotein. J Cell Biol 103:255.
5. Tykocinski ML, Shu K-K, Ayers DJ, Walter EI, Getty RR, Groger RK, Hauer CA, Medof ME (1988). Glycolipid re-anchoring of T-lymphocyte surface antigen CD8 using the 3' end sequence of decay-accelerating factor's mRNA. Proc Natl Acad Sci USA 85:3555.
6. Caras IW, Weddell GN, Davitz MA, Nussenzweig V, Martin DW (1987). Signal for attachment of a phospholipid membrane anchor in decay accelerating factor. Science 238:1280.
7. Caras IW, Weddell GN (1989). Signal peptide for protein excretion directing glycophospholipid membrane anchor attachment. Science 243:1196.
8. Waneck GL, Sherman DH, Kincade PW, Low MG, Flavell RA (1988). Molecular mapping of signals in the Qa-2 antigen required for attachment of the phosphatidylinositol membrane anchor. Proc Natl Acad Sci USA 85:577.
9. Tabor S, Richardson CC (1987). DNA sequence analysis with a modified bacteriophage T7 DNA polymerase. Proc Natl Acad Sci USA 84:4767.

SURVIVAL OF LEISHMANIA PARASITES WITHIN PHAGOCYTIC CELLS: REQUIREMENT FOR LIPOPHOSPHOGLYCAN[1]

S. J. Turco, D. L. McLean, and T. B. McNeely

Department of Biochemistry, University of Kentucky
College of Medicine, Lexington, KY 40536

ABSTRACT Lipophosphoglycan is the major cell surface glycoconjugate of Leishmania parasites. A striking characteristic of these parasites is their ability to avoid destruction within phagolysosomes of host phagocytic cells. We have proposed that lipophosphoglycan is critical in enabling the parasite to survive in hostile environments. Consistent with this hypothesis, the incubation of human peripheral monocytes with a variant line of L. donovani parasites which lack lipophosphoglycan resulted in the entry of the variant cells into the monocytes and their subsequent destruction, which was in contrast to wildtype L. donovani.

INTRODUCTION

The protozoan parasite Leishmania donovani is the causative agent of the often fatal disease kala-azar. In its digenetic life cycle, L. donovani avoids destruction in two hydrolytic environments. The parasites reside extracellularly in the alimentary tract of their sandfly vector and, upon infection of humans, they proliferate in the phagolysosomes of cells of the reticuloendothelial

---

[1]This work was supported by NIH grants AI20941 and an UNDP/World Bank/WHO Special Programme for Research and Training in Tropical Diseases. S.J.T. is a Burroughs Wellcome Scholar in Molecular Parasitology.

system. The major cell surface glycoconjugate of L. donovani is lipophosphoglycan (LPG), which possibly plays a significant role in the survival of the parasites in these hostile environments (1-3). Structurally, LPG is a heterogeneous polymer of repeating [PO$_4 \rightarrow$Gal(β 1,4)Manα1] disaccharide units attached via a phosphosaccharide core to a novel lyso-alkylphosphatidylinositol anchor (4,5). A similar, but immunologically distinct glycoconjugate has been reported in all species of Leishmania (6-9).

In this article, evidence is provided establishing the importance of LPG in enabling the parasite to survive within phagocytic cells. The molecular basis for the role of LPG is discussed.

MATERIALS AND METHODS

Cells and Cell Culture

Leishmania donovani parasites were passaged in Syrian hamsters. Hamsters (3-4 wks old) were infected intraperitoneally with 10^8 promastigotes. After 8-12 weeks, hamsters were sacrificed. Infected spleens were removed and homogenized. Amastigotes were isolated by differential centrifugation and red blood cells were removed by treatment with saponin. The purified amastigotes were then reinjected into hamsters or allowed to differentiate into promastigotes in d-DME media. Promastigotes were cultured according to Iovannisci and Ullman (10). The cells were grown at 25°C in Dulbecco's modified Eagle's medium (DME) supplemented with 0.3% bovine serum albumin, 10% fetal calf serum, adenosine (0.05 mM), xanthine (0.05 mM), biotin (1 mg/l), Tween 80 (40 mg/l), hemin (5 mg/l), and triethanolamine (0.5 ml/l). This supplemented medium will be referred to as d-DME.

Mutagenesis and Cloning of Parasites

Mutagenesis of the parasites and cloning of the RT5 ricin-resistant clone was conducted according to the procedure of King and Turco (11). Briefly, 300 ml of wildtype cells at a density of 10^6 cells/ml were incubated with the mutagen N-methyl-N-nitroso-N'-nitroguanidine at a concentration of 4 µg/ml in d-DME for 3.5 h. The parasites were centrifuged at 3000xg for 10 min, resuspended in fresh d-DME, and passaged for several generations before use.

Cells were cloned on soft agar as follows: equivalent amounts of 2X d-DME (twice the normal concentration of ingredients) and sterile, liquified agar (2%) were mixed. A sterile preparation of ricin agglutinin was added to give a final concentration of the lectin of 100 µg/ml. The mixture (25 ml) was pipetted onto 100 mm petri dishes and allowed to solidify for at least 7 h. For cloning of parasites, approximately 80 µl of concentrated cells in d-DME were streaked onto the soft agar dishes (11).

Metabolic Labeling and Extraction

Exponentially growing cells ($2x10^6$-$2x10^7$ cells/ml) were metabolically-labeled with radioactive mannose or galactose and extracted as described previously (1). The solvent water/ethanol/diethylether/pyridine/NH_4OH (15:15:5:1:0.017) was used to solubilize LPG during the extraction protocol.

Isolation of Human Monocytes

Human monocytes were isolated from 100ml of fresh blood collected into EDTA. Sepracell-MN^R, a colloidal silica mixture, was used to isolate monocytes according to the manufacturer's specifications. Briefly, equal volumes of anticoagulated blood and Sepracell-MN were mixed and centrifuged at 1,500xg for 20 minutes at 23°C in a swinging bucket rotor. The top 20% of the gradient containing the monocytes (among other cells) was removed and washed 1X with PBS containing 0.1% BSA (PBS-BSA). The pellet from the wash was resuspended in a 2:3 (v/v) mix of PBS-BSA and Sepracell-MN to give a total volume 1/2 of the volume of the first gradient. This mix was centrifuged at 1,500xg for 20 minutes as above. The top 20% of this gradient containing mainly monocytes was removed and washed 2X with PBS-BSA. This pellet was centrifuged in 5 ml of PBS-BSA at 1000xg for 5 min to remove platelets. The pellet was then resuspended in 10 ml of media 199 containing 10% heat-denatured fetal calf serum, 10mM Hepes and 100U/ml penicillin and streptomycin. Cells were cultured in this media with 5% CO_2 at 37°C until use. Cells were used from day 0 to day 4 of culture.

RESULTS

Selection of the RT5 Ricin-resistant Clone of L. donovani

Mutagenized L. donovani (300 ml of 5×10^6 cells/ml) were centrifuged at 3000xg for 10 min and resuspended in 10 ml of fresh d-DME. Ricin was added to a final concentration of 100 μg/ml for 2 hr during which massive agglutination of the cells occurred. The agglutinated cells were removed by centrifugation at 500xg for 10 min. Cells remaining in the medium were pelleted at 3000xg for 10 min and incubated in 10 ml of fresh d-DME containing 100 μg/ml of ricin for 3 days. Any agglutinated cells were removed by centrifugation at 500xg and nonagglutinated cells were pelleted at 3000xg for 10 min. The latter were streaked onto 100 mm petri dishes containing 100 μg/ml of ricin in agar. Several distinct colonies were isolated two weeks after plating. One of these, RT5, was characterized further.

Characterization of the RT5 Line

RT5 cells were similar in all respects to another ricin-resistant L. donovani described elsewhere (11): (i) a generation time of 16-18 h (comparable to that of wildtype parental cells), (ii) approximately 100-fold more resistant to the cytotoxic effects of ricin, (iii) approximately 10-fold more sensitive to the cytotoxic effects of concanavalin A, and (iv) totally deficient in the synthesis and expression of LPG as judged by metabolic labeling with either [^{3}H]galactose or [^{3}H]mannose (Table 1).

Infection of WT and RT5 Parasites in Monocytes

One day old human peripheral monocytes were incubated with WT and RT5 parasites for 4 h in a ratio of 10:1 parasites to monocytes. The monocytes were then washed free of any extracellular parasites and were further incubated. The number of intracellular parasites per monocyte was quantitated. As shown in Fig. 1, the average number of WT parasites per monocyte declined within several days and then

TABLE 1

Incorporation of [^{3}H]Mannose and [^{3}H]Galactose into Glycoproteins and LPG

	[^{3}H]Man Incorporation		[^{3}H]Gal Incorporation	
	A	B	A	B
WT	5,000	2,000	5,400	1,100
RT5	2,200	60	2,600	50

Parasites were metabolically-labeled with 100 µCi of either [^{3}H]mannose or [^{3}H]galactose for 2 h, extracted as described previously (11), and measured for radioactivity (cpm) incorporated into the various complex carbohydrates. A: Glycoproteins; B: LPG.

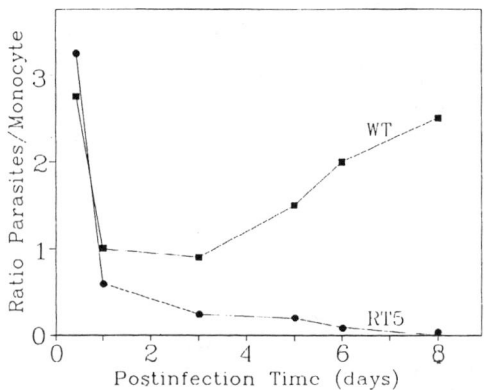

FIGURE 1. Growth of WT and RT5 within human peripheral monocytes. Monocytes (at day 1 of culture) were infected with promastigotes (at day 6 of growth) with a parasite to monocyte ratio of 10:1. The infection was carried out in media 199 containing 10% heat-denatured fetal calf serum, at 37°C with rocking, for 4 hours. The cells were pelleted by centrifugation and resuspended in media 199 containing 10% fresh autologous serum (all non-phagocytized parasites were killed in this media). At designated times, 100 µl of infected cells were removed and cytospun onto microscope slides. Slides were stained with Diff-Quik stain and the number of intracellular amastigotes was quantitated.

the WT parasites began to multiply. A similar observation was previously reported for L. donovani (12). RT5 parasites easily entered the monocytes, but the number decreased after several days and eventually all intracellular RT5 parasites were destroyed.

Attenuation of the Oxidative Burst in Monocytes by LPG

To explore a potential basis for the role of LPG, purified LPG was examined for its ability to influence production of the oxidative burst in human monocytes. Monocytes were preincubated with 50 µM LPG for 10 min. The monocytes were then stimulated for the production of the oxidative burst with either phorbol myristic acid (PMA) or with opsonized zymosan. The oxidative burst was quantitated with an assay that involved chemiluminescence (Fig. 2).

In the upper panel of Fig. 2, treatment of the monocytes with PMA in the absence of LPG resulted in a maximal stimulation of the oxidative burst within 10 min and a gradual decline in oxidative burst activity. Pretreatment of the monocytes with 50 µM LPG resulted in a 30-40% attenuation of the burst. Shown in the lower panel is a similar experiment with zymosan. In the absence of LPG, there was a delayed stimulation of the burst which reached a maximum in 25 min. Pretreatment of the monocytes with LPG resulted in 70% attenuation of oxidative burst activity.

Inhibitory Effect of LPG on Protein Kinase C

Induction of the oxidative burst in activated phagocytic cells is believed to be mediated by protein kinase C. To examine the hypothesis that the inhibition of activation of the oxidative burst by L. donovani might be accomplished by inhibition of protein kinase C by LPG, the enzyme was purified from rat brains according to the procedure of Walton et al. (13). Protein kinase C was assayed by its ability to phosphorylated histones using $[\gamma-^{32}P]ATP$ as the phosphate donor. In addition, LPG was fragmented by several techniques (Fig. 3) and the fragments were purified and examined for inhibitory activity toward protein kinase C.

As reported earlier (2), intact LPG is a potent inhibitor of the enzyme even at a relatively high concentration of phosphatidylserine (36 µM). The

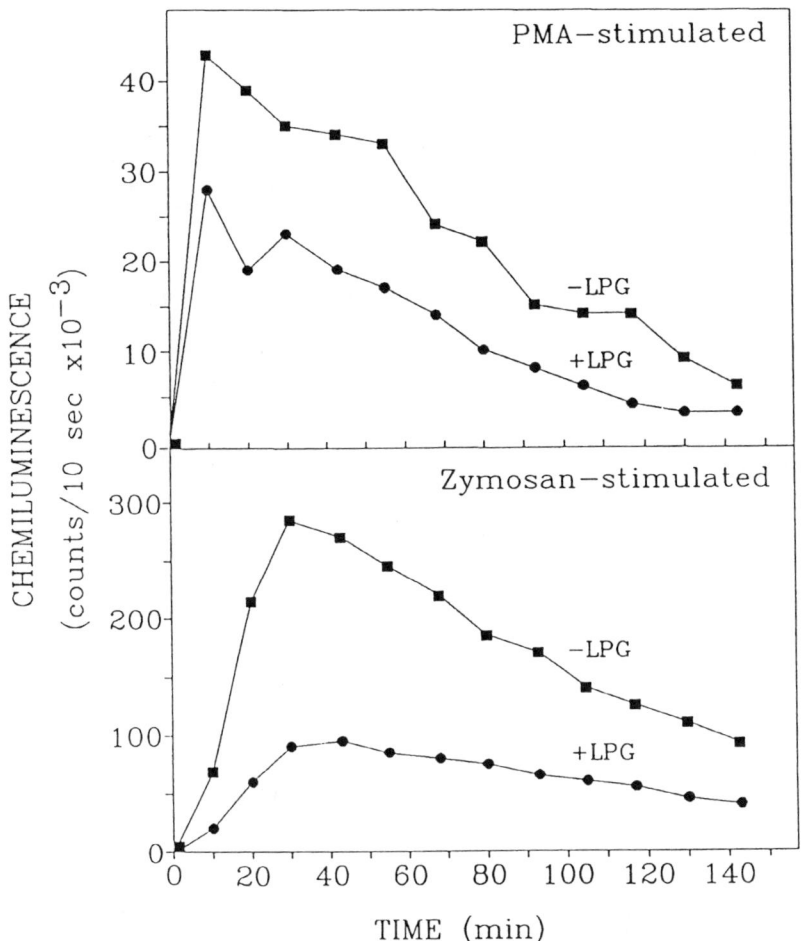

FIGURE 2. Attenuation of PMA/Zymosan-stimulated Oxidative Burst in Human Monocytes by LPG. 50 µM LPG in PBS or PBS alone was added to 1×10^5 monocytes (at day 1 of culture) in 100 µl of media 199 containing 10% heat-denatured fetal calf serum and 4×10^{-5}M luminol. Monocytes were incubated with the LPG/PBS for 10 minutes at 25°C and then stimulated by the addition of either 10 µM PMA or 1.0 mg/ml zymosan opsonized with autologous serum. The oxygen burst was followed by monitoring the fluorescence of luminol in a scintillation counter. The samples were counted repeatedly for 10 seconds over a period of 150 minutes.

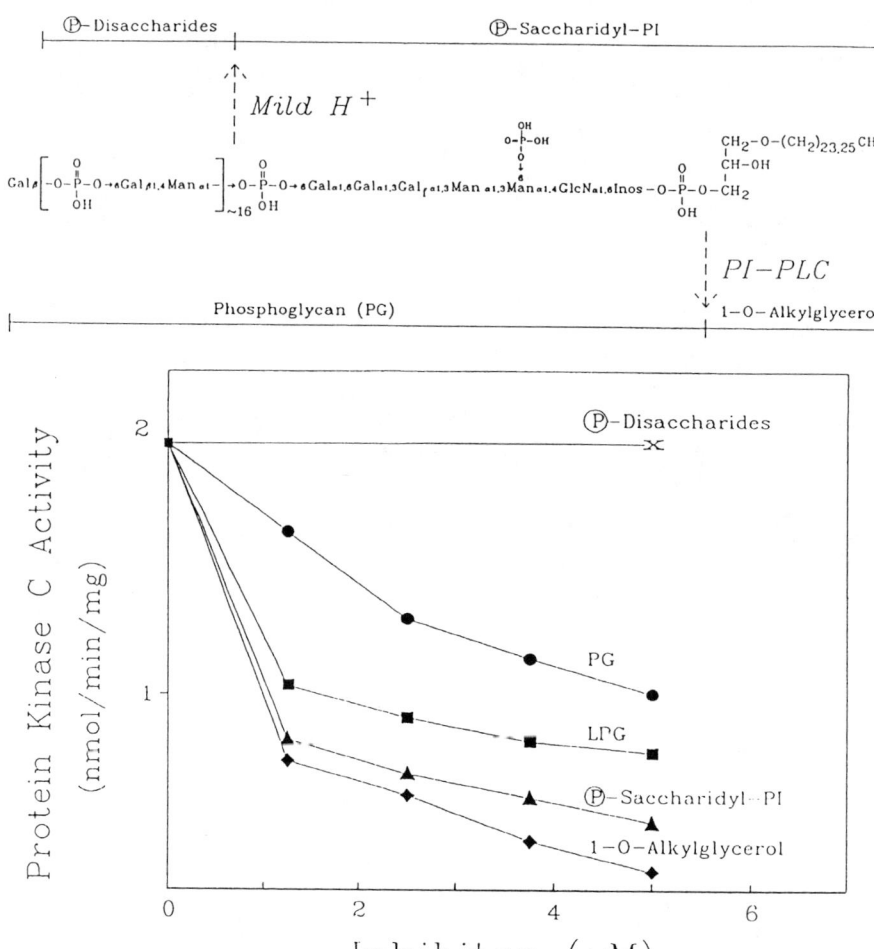

FIGURE 3. Fragmentation scheme for LPG and inhibition of protein kinase C activity by LPG and LPG fragments. Purified LPG was treated with 0.02 $\underline{N}$ HCl, 5 min, 100°C to generate the phosphorylated disaccharides and phosphosaccharidyl-PI portion or with the enzyme PI-PLC to obtain the 1-O-alkylglycerol and the entire carbohydrate portion termed phosphoglycan. Protein kinase C was assayed in the presence of 1.0 mM $CaCl_2$, 5.0 mM $MgCl_2$, 36 µM phosphatidylinositol, and 0.15 µM diacylglycerol with the designated concentrations of LPG and LPG fragments.

1-O-alkylglycerol fragment of LPG (4), which consists of either a saturated, unbranched C_{24} (78%) or C_{26} (22%) hydrocarbon, had a greater inhibitory effect on protein kinase C than intact LPG under these assay conditions (Fig. 3). The phosphosaccharidyl-PI fragment also had a greater inhibitory effect than intact LPG. Although the entire carbohydrate portion of LPG (phosphoglycan) also suppressed the activity of protein kinase C, it was not as effective as LPG or its lipid-containing fragments. Addition of the individual phosphorylated disaccharide fragments isolated from LPG had no effect on kinase activity.

DISCUSSION

Since LPG is the major surface glycoconjugate of leishmanial parasites, it is very possible that LPG plays a critical role in the survival of the parasites within host cells. The Leishmania donovani variant RT5 was used to establish the importance of LPG in enabling the parasite to survive within human peripheral monocytes. RT5 parasites were selected for their resistance to ricin agglutinin and were shown to be totally lacking in the synthesis and expression of LPG. The variant cells were examined for their ability to enter and sustain an infection in the monocytes. The results indicated that the variant cells easily entered the phagocytic cells, but were unable to survive and were destroyed several days postinfection. It is concluded that LPG is indeed crucial in the overall infection process.

The molecular basis for the role of LPG in this process is currently being investigated. One important survival mechanism would be the potential abrogation or mediation of the microbicidal oxidative burst of the phagocytic cells in which the parasites reside. Our results have shown that exogenously added LPG attenuates the stimulation of the oxidative burst when monocytes are induced with either phorbol myristic acid or opsonized zymosan. There are several possibilities that could explain the observed attentuation effect by LPG. An artifactual reduction of chemiluminescence by LPG on the luminol-based assay was eliminated by showing that LPG had no effect on the luminol reaction in the presence of excess hydrogen peroxide (data not shown). Additionally, LPG was found not to be cytotoxic to monocytes with the concentrations used (0-100 µM). In a significant possible explanation, it is conceivable that LPG serves as a scavenger of one or more of the elements of the oxidative burst (i.e. superoxide anion, hydrogen peroxide, hydroxyl radical, etc.). We are pursuing this aspect.

Another possibility in which some evidence is available is that LPG might prevent induction of the normally dormant oxidative burst. Induction of the burst is believed to be mediated by protein kinase C (14-16). Recently, we demonstrated that the <u>L. donovani</u> LPG is a potent <u>in vitro</u> inhibitor of purified protein kinase C (2). The glycoconjugate is a competitive inhibitor with respect to diolein and a noncompetitive inhibitor with respect to phosphatidylserine. Importantly, LPG has no effect on protein kinase M (believed to be the catalytic domain of protein kinase C and cannot be stimulated by phosphatidyl serine, diolein, and calcium) and protein kinase A (2). The results from this report indicate that the 1-O-alkylglycerol portion of LPG is primarily responsible for the inhibitory activity on protein kinase C. This was not unexpected since ether-for-acyl substitutions have been reported to result in a potent reduction in protein kinase C-activating ability (17,18). The type of inhibition displayed by the 1-O-alkylglycerol appears to be complex with respect to diolein concentration since both the apparent Km and Vmax were affected (data not shown). A similar type of inhibition has been previously reported to occur in the presence of alkylacylglycerols (19).

A potentially important observation concerning LPG that actually accounted for its original name of "excreted factor" (20) is its release from the surface of the parasite and its appearance in the culture medium. One of the release forms of LPG is a hydrophilic form (1,20). While the identity of the cleavage reaction has not yet been demonstrated, the resulting hydrophobic fragment may actually be the biological inhibitor of protein kinase C within infected macrophages.

In another potential mechanism of inhibition, it is possible that LPG (or PG), due to the presence of many charged phosphate groups, could chelate significant quantities of free calcium, or other important divalent cations, which are present in small amounts intracellularly. Indeed, when assayed in the presence of micromolar levels of calcium, the difference in inhibitory activity between LPG and 1-0-alkylglycerol becomes almost negligible, perhaps due, in part, to chelation of the calcium (data not shown).

ACKNOWLEDGEMENTS

The authors are indebted to Drs. Charles J. Waechter and Robert L. Lester for many helpful discussions during the course of this work.

REFERENCES

1. King, D. L., Chang, Y. D., and Turco, S. J. (1987) Mol. Biochem. Parasitol. 24, 47-54.
2. McNeely, T. B. and Turco, S. J. (1987) Biochem. Biophys. Res. Commun. 148, 653-657.
3. Turco, S. J. (1988) Biochem. Soc. Trans. 16, 259-261.
4. Orlandi, P. A. and Turco, S. J. (1987) J. Biol. Chem. 262, 10384-10391.
5. Turco, S. J., Hull, S. R., Orlandi, P. A., Shepherd, S. D., Homans, S. W., Dwek, R. A, and Rademacher, T. W. (1987) Biochemistry 26, 6233-6238.
6. Hernandez, A. G. (1983) in Cytopathology of Parasitic Diseases, CIBA Foundation Symposium 99, Pitmans, London, pp. 138-156.
7. Handman, E., Greenblatt, C. L., and Goding, J. W. (1984) EMBO J. 3, 2301-2306.
8. Handman, E. and Goding, J. W. (1984) EMBO J. 4, 329-336.
9. McConville, M. J., Bacic, A., Mitchell, G. F., and Handman, E. (1987) Proc. Natl. Acad. Sci. USA 84, 8941-8945.
10. Iovannisci, D. M., and Ullman, B. (1983) J. Parasitol. 69, 633-636.
11. King, S. J. and Turco, S. J. (1988) Mol. Biochem. Parasitol. 28, 285-294.
12. Pearson, R. D., Harcus, J. L., Symes, P. H., Romito, R., and Donowitz, G. R. (1982) J. Immunol. 129, 1282-1286.
13. Walton, G. M., Bertics, P. J., Hudson, L. G., Vedvick, T. S., and Gill, G. N. (1987) Anal. Biochem. 161, 425-437.
14. Wilson, E., Olcott, M. C., Bell, R. M., Merrill, A. H., and Lambeth, J. (1986) J. Biol. Chem. 261, 12616-12623.

15. Pontremoli, S., Melloni, E., Salamino, F., Sparatore, B., Michetti, M., Sacco, O., and Horecker, B. L. (1986) Biochem. Biophys. Res. Commun. 140, 1121-1126.
16. Gennaro, R., Florio, C., and Romeo, D. (1985) FEBS Lett. 180, 185-190.
17. Ganong, B. R., Loomis, C. R., Hannun, Y. A., Bell, R. A. (1986) Proc. Natl. Acad. Sci. USA 83, 1184-1188.
18. Heymanns, F. D., DaSilva, C., Marrec, N., Godfroid, J. J., Castagna, M. (1987) FEBS Lett. 218, 35-40.
19. Daniel, L. W., Small, G. W., and Schmitt, J. D. (1988) Biochem. Biophys. Res. Commun. 151, 291-297.
20. Slutzky, G. M., El-On, J., and Greenblatt, C. L. (1979) Infect. and Immunol. 26, 916-924.

POST-TRANSLATIONAL MODIFICATIONS OF THE CELLULAR AND SCRAPIE PRION PROTEINS[1]

Stanley B. Prusiner

Departments of Neurology and of Biochemistry and Biophysics
University of California
San Francisco, California 94143-0518

ABSTRACT Prions are novel, transmissible pathogens causing degenerative diseases in humans and animals. Kuru, Creutzfeldt-Jakob disease (CJD) and Gerstmann-Sträussler syndrome (GSS) illustrate the infectious, sporadic and genetic mechanisms for human prion diseases, respectively. Scrapie of sheep and goats is the prototypic prion disorder since it was the first of these diseases to be transmitted to laboratory rodents. Prions are composed largely, if not entirely, of an abnormal isoform of the prion protein (PrP). Monoclonal antibodies (mAb) raised against PrP 27-30, which is derived from the scrapie PrP isoform (PrP^{Sc}) by limited proteolysis, have been used to purify scrapie prion infectivity in detergent-lipid-protein complexes (DLPC). Immunoaffinity-purified fractions contain PrP^{Sc} and high prion titers. Polyclonal antibodies to PrP 27-30 were found to neutralize scrapie infectivity. To date, no prion-specific nucleic acid has been identified which is required for transmission of disease. PrP^C and PrP^{Sc} are thought to have the same amino acid sequence but differ due to some post-translational process. Both PrP^C and PrP^{Sc} are glycoproteins which possess Asn-linked oligosaccharides and glycosyl phosphatidylinositol (GPI) anchors. Whether the

[1]This work was supported by research grants from the National Institutes of Health (AG02132, NS14069), a Senator Jacob Javits Center of Excellence in Neuroscience (NS11786) and a contract from the California State Department of Health Services (87-92062) as well as by gifts from Sherman Fairchild Foundation and RJR/Nabisco, Inc.

features which distinguish PrP^{Sc} from PrP^C arise from differences in their Asn-linked oligosaccharides or GPI anchors is unknown. Variations in PrP amino sequence are linked genetically to altered prion disease phenotypes in both humans and mice. Much recent experimental data have established that prions are novel and unprecedented pathogens. Indeed, the uniqueness of prions promises to open many new and unexpected avenues of research.

INTRODUCTION

Prions continue to present an unparalleled opportunity to decipher some of the pathogenic mechanisms responsible for brain degeneration. The study of prion diseases has occupied a rather romantic yet most perplexing venue in the biomedical sciences for nearly three decades. The exotic but primitive culture of the kuru region in the Eastern Highlands of Papua New Guinea stimulated interest in the nature of the infectious pathogens transmitted by ritualistic cannibalism (1), while the unusual molecular properties of the prion particle created considerable attention for two rare human disorders found all over our planet: CJD and GSS.

Besides the three human prion diseases noted above, four prion diseases of animals are now recognized. These are scrapie of sheep and goats, transmissible mink encephalopathy (2), chronic wasting disease of mule deer and elk (3), and bovine spongiform encephalopathy (4, 5). Of all these transmissible degenerative diseases of the central nervous system in humans and animals, scrapie has been studied most extensively.

PRIONS CONTAIN PrP^{Sc}

PrP 27-30 was discovered by enriching brain fractions for scrapie infectivity (6, 7). Development of a more rapid and economical bioassay (8) greatly facilitated purification of the hamster scrapie agent (6, 7). PrP 27-30 migrates during sodium dodecyl sulfate-polyacrylamide gel electrophoresis (SDS-PAGE) as a broad band with an apparent molecular weight (M_r) of 27,000 to 30,000.

Ten lines of evidence establish that the protein PrP^{Sc} (or PrP 27-30) is a component of the infectious prion particle: (a) PrP 27-30 and the scrapie agent copurify (7, 9, 10) using detergent extractions and limited proteolysis to promote aggregation of prions into amyloid rods which are collected by centrifugation. PrP 27-30 is the most abundant macromolecule in purified preparations (10). (b) Immunoaffinity purification of scrapie prions was accomplished with PrP mAb coupled to protein A-Sepharose. Dispersion of prions into DLPC was required before meaningful chromatography could be performed (11). Copurification of scrapie prion infectivity and PrP^{Sc} was found in fractions eluted from a PrP mAb column (12). Indeed, copurification by two different procedures argues that the molecular structure and properties of of PrP^{Sc} (as well as PrP 27-30) and the infectious prion particles must be extremely similar. (c) The PrP 27-30 concentration is proportional to the prion titer (13). PrP^{Sc} is absent in normal, uninfected animals (14, 15). (d) Rabbit antisera raised against PrP 27-30 purified by SDS-PAGE was found to neutralize scrapie infectivity in DLPC (12). However, no neutralization of prion infectivity was observed with prions aggregated in amyloid rods (16, 17). (e) Procedures that denature, hydrolyze, or selectively modify PrP 27-30 also diminish the prion titer (13). The unusual kinetics of PrP 27-30 hydrolysis catalyzed by proteases were found to correlate with the diminution of prion titer. (f) PrP 27-30 and scrapie infectivity copartition into many different forms — membranes, rods, spheres, DLPC, and liposomes. These dramatically different physical forms all contain PrP 27-30 and high prion titers (7, 9, 10, 18-21). (g) The PrP gene (Prn-p) in mice is linked to a gene controlling scrapie-incubation times (Prn-i) (22). Prolonged incubation periods are a cardinal feature of both scrapie and CJD. The preeminent role of PrP in the pathogenesis of scrapie has been made more compelling by the discovery of a correlation between PrP amino acid sequence and scrapie-incubation times (23). (h) An amino acid substitution in the human PrP gene is linked genetically to the development of GSS (24, 25). GSS and familial CJD are the only known human diseases which are both genetic and infectious. (i) Scrapie and CJD prion proteins have been identified only in tissues of animals and humans with transmissible neurodegenerative diseases and not in those with other disorders, such as murine systemic amyloidosis and human Alzheimer's disease, anoxic encephalopathy, or non-neurologic disorders (7, 10, 26-32). (j) Cultured murine neuroblastoma cells have been infected with both scrapie and CJD prions (33-35). Clones of the scrapie-infected cells were found to produce PrP^{Sc}, whereas clones showing no infectivity lacked PrP^{Sc} (33).

Many investigators have confirmed the presence of PrP 27-30 in brains infected with the scrapie or CJD agent (30, 36-38). Although the amino acid sequence of PrP has been confirmed and there is agreement that PrP is glycosylated, some investigators have suggested that PrP 27-30 may not be a component of the scrapie agent (39, 40). One argument revolves around the inability of some investigators to detect PrP mRNA or PrP^{Sc} in spleens of scrapie-infected rodents (41, 42); however, investigators in several laboratories have clearly shown that both PrP mRNA and PrP^{Sc} are present in spleen tissue (15, 43-45). Indeed, recent studies have shown an excellent correlation between the concentration of PrP mRNA in scrapie-infected rodent tissues and prion titer (43). Another argument centers on experiments that demonstrate the loss of CJD infectivity in fractions following lectin chromatography. Neither denaturation of the PrP CJD nor neutralization of CJD infectivity by immobilization on the lectin column was considered as an explanation (38). Denaturation of PrP^{Sc} has been demonstrated to be accompanied by a loss of scrapie infectivity (9, 10). Indeed, no experimental studies have been reported where fractions with high levels of scrapie infectivity were found to contain less than one PrP^{Sc} (or PrP 27-30) molecule per infectious unit.

STRUCTURE AND ORGANIZATION OF PrP GENES

Once the N-terminal amino acid sequence of PrP 27-30 was determined (9), oligonucleotides corresponding to a portion of this sequence were synthesized and used to identify a PrP cDNA (15, 41). Southern blotting with PrP cDNA revealed a single-copy gene with the same restriction patterns as those in normal and scrapie-infected DNA from hamster brains. Unexpectedly, PrP mRNA was found at similar levels in both normal and scrapie-infected hamster brains.

The organization and structure of the hamster PrP gene have been elucidated; the entire open reading fram (ORF) or protein coding region is contained within a single exon (46). The 5' end of the PrP gene contains multiple G:C-rich initiation sites. An intron of ~10 kb separates exons I and II. Six lines of evidence argue that the two PrP isoforms have the same amino acid sequence: (a) No evidence for rearrangement of the PrP gene in scrapie has been found (15,

46). (b) The organization of the PrP gene provides no possibility for alternative splicing within the ORF (12). (c) Only one PrP mRNA of 2.1 kb has been detected and its concentration does not change throughout the course of scrapie infection in hamster brain (15). (d) 70% of PrP^{Sc} and 87% of PrP 27-30 have been sequenced by gas-phase protein sequencing (D. Teplow, D. Groth, L. Hood, S. Prusiner, unpublished data). These sequences correspond precisely with the translated genomic DNA sequence. (e) Both scrapie hamster brain and mouse PrP cDNAs have translated ORF sequences which are identical with the corresponding translated genomic sequences (15, 23, 46, 47). (f) Both PrP isoforms have the same N-terminal amino acid sequence as determined by N-terminal gas-phase sequencing (48). Presumably, the difference in the properties of the two prion proteins is due to a post-translational event (46); though seemingly remote, the possibility that differences in the amino acid sequences of the two PrP isoforms arising through RNA editing must continue to be considered (49-52).

The human and mouse PrP genes have been shown to be located on chromosomes 20 and 2, respectively, which are homologous (53). This finding indicates that PrP genes existed prior to the speciation of mammals. The ORF of PrP genes from humans (54), Syrian hamsters (46), Chinese hamsters (55), Armenian hamsters (55), mice (23, 47), rats (56) and sheep (57) have been sequenced and all encode prion proteins of approximately 250 amino acids with N-terminal signal peptides. All of these PrP ORFs encode C-terminal hydrophobic peptides which are presumably removed upon GPI anchor addition and all contain two concensus sites for Asn-linked glycosylation as well as two cysteines within the C-terminal half of the molecule. All of the ORFs also possess a series of Gly-rich repeats in the N-terminal portion of the PrP molecule. Whether eukaryotes other than mammals have authentic PrP genes remains to be established (58).

PRION PROTEIN ISOFORMS

PrP antibodies detected proteins of M_r 33-35 kDa in both normal and scrapie-infected brains (14, 15). PrP^C is degraded by proteinase K digestion while the N-terminal 67 amino acids of PrP^{Sc} are removed to form PrP 27-30 (Figure 1; 15, 21, 59). Both PrP^C and PrP^{Sc} are membrane proteins, but upon detergent extraction PrP^C is solubilized, whereas PrP^{Sc} after limited proteolysis polymerizes into amyloid rods. This difference in solubility allows separation of the two PrP isoforms by centrifugation (60).

That protease resistance of PrP 27-30 and prion infectivity is not due to aggregation alone was demonstrated by studies with DLPC. Similar inactivation kinetics for prions in rods and DLPC were observed with proteinase K digestions (11) while PrP 27-30 in rods could be chemically crosslinked but PrP 27-30 in DLPC could not (R. Gabizon, D. Groth, S. Prusiner, in preparation). Thus, the protease resistance, at least in part, of PrP 27-30 appears to be independent of its aggregation state.

Protease-resistant prion proteins have been found in the brains of patients dying of CJD, GSS and kuru (26, 27, 61). Rod-shaped amyloids have been found in fractions purified from CJD brains (26), suggesting that amyloid plaques in the brains of CJD, GSS and kuru patients are composed of paracrystalline arrays of prion proteins (31, 32).

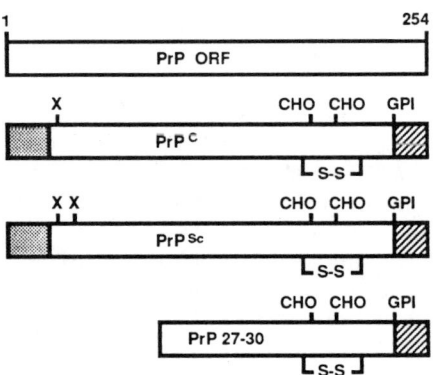

FIGURE 1. Prion protein structure. Hamster PrP gene encodes a protein of 254 amino acids. The N-terminal signal peptide of 22 amino acids (stippled pattern) is cleaved during maturation of PrP^C and PrP^{Sc}. Modified Arg residues (X) at codons 25 and 37 have been identified as well as Asn-linked oligosaccharides (CHO) attached at codons 181 and 197 within a loop formed by a disulfide bond which joins Cys at codons 179 and 214. Upon removal of a C-terminal hydrophobic peptide of ~20 amino acids (diagonal line pattern), a glycosyl phosphatidylinositol (GPI) anchor is added.

POST-TRANSLATIONAL MODIFICATIONS

At present, there is evidence for eight post-translational modifications of PrP (Figure 1). The first modification to be identified was glycosylation; PrP 27-30 is a sialoglycoprotein (6, 9). All translated PrP gene ORFs sequenced to date contain two consensus sites for Asn-linked glycosylation. In Syrian hamster PrP, these consensus sites are codons 181 and 197 (15, 46). Both the cellular and scrapie PrP isoforms are resistant to digestion by endoglycosidase H but possess N-linked oligosaccharides that can be removed by digestion with peptide:N-glycosidase F (62). Hydrazinolysis liberated ~2 moles of oligosaccharides per mole of PrP 27-30 (T. Endo, D. Groth, S. Prusiner, A. Kobata, submitted for publication). The released oligosaccharides were found to be a mixture of bi-, tri-, and tetra-antennary complex-type sugar chains with Manα1→6 (GlcNAcβ1→4) (Manα1→3) Manβ1→4GlcNAcβ1→4 (Fucα1→6) GlcNAc as their cores. Variation is produced by combinations of the following oligosaccharides Galβ1→4GlcNAcβ1→, Galβ1→4 (Fucα1→3) GlcNAcβ1→, GlcNAcβ1→, Neu5Acα2→3Galβ1→4GlcNAcβ1 and Neu5Acα2→6Galβ1→4GlcNAcβ1→ in their outer chains. Whether the structures of the Asn-linked oligosaccharides attached to PrP^C differ significantly from those of PrP^{Sc} remains to be determined.

Both mature PrP isoforms have two cysteine residues at codons 179 and 214 that are linked by an intramolecular disulfide bond (48). This disulfide bond forms a loop which contains both Asn-linked oligosaccharides as well as a variant amino acid at codon 189 found in long incubation period mice (Prn-p^b).

The N-terminus of PrP contains a 22-amino-acid signal peptide. Like all other signal peptides, it contains a hydrophobic core and a consensus cleavage site (46, 63). Cell-free translation studies have demonstrated the cleavage of the PrP signal peptide (64) and N-terminal amino acid sequencing of PrP^C and PrP^{Sc} confirm the cleavage (37, 48, 65).

Like the N-terminal signal peptide (37, 46, 64, 66), the C-terminal hydrophobic segment of both PrP isoforms appears to be removed during maturation (15) and a GPI anchor added (67). PrP^C is localized almost exclusively on the external surface of cultured neuronal cells; the topology of PrP^{Sc} remains to be established.

Interestingly, PrP^{Sc} is resistant to release from cell membranes by phosphatidylinositol-specific phospholipase (PIPLC) digestion while PrP^C is quantitatively released from the surface of cultured cells (68). Studies with scrapie-infected murine neuroblastoma cells suggest that PrP^{Sc} remains largely within the interior of these cells while PrP^C is exported to the external cell surface (69).

Arg_{25} and Arg_{37} in PrP^{Sc} contain an unknown modification which prevents their detection by gas-phase sequencing using the Edman degradation (48, 70). Arg_{25} of PrP^C contains a similar modification (48). Whether Arg_{37} also contains this modification is uncertain; however Arg_{48} of PrP^{Sc} does not contain the modification.

ON THE FUNCTION OF PrP^C

The localization of PrP^C to the external surface of cells, developmental regulation of PrP^C expression and the x-antigenic determinant (Galβ1→4 [Fucα1→3] GlcNAc) within the outer chains of its Asn-linked oligosaccharides all suggest a role in cell recognition. In support of the hypothesis that PrP^C may be involved in cell recognition are studies with rat neuroblasts showing greatly increased levels of PrP^C on the surface under conditions where they differentiate into neurons (71). Cultured rat neuroblasts and glial cells transformed with a temperature-sensitive (ts) Rous sarcoma virus (RSV) (72) exhibit low or nondetectable levels of PrP^C on their surface at the permissive temperature (34°C) but greatly increased PrP^C levels at the nonpermissive temperature (38°C) where they differentiate.

Three different putative roles for PrP^C in scrapie have been identified. First, PrP^C may prevent an immune response to PrP^{Sc} and thus explain one of the most puzzling questions in research on scrapie and CJD. PrP^C may account for the lack of an immune response to a lethal "slow infection" by rendering the host tolerant to the abnormal isoform (PrP^{Sc}) (15, 73). The difficulties in raising antibodies to PrP 27-30 may be at least partly due to tolerance induced by PrP^C (14). Second, PrP^C may modulate the conversion of PrP^C or a precursor into PrP^{Sc}. This hypothesis is consistent with the observations that long incubation periods are dominant in (Prn^a x Prn^b)F1 mice (22, 23, 74-78), that heterologous recombinant PrP expressed in scrapie-infected cultured cells inhibits production of PrP^{Sc} (M. Scott, S. Prusiner, in preparation) and that a mutation in one allele of the human

PrP gene is linked to the development of GSS (24). Third, PrP^C may modulate the initiation of scrapie infection. This hypothesis is supported by recent findings that isologous prions produce scrapie more rapidly in Prn^a and Prn^b mice (79) as well as by the discovery of a reduction in incubation times in neonatal hamsters whose brains lack PrP^C (80, 81).

If PrP^C has a role in modulating prion infections, one model would predict the binding of PrP^C to PrP^{Sc}. However, attempts to demonstrate binding of purified PrP 27-30 to PrP^C using photoactivatable crosslinking reagents have been unsuccessful (R. Gabizon and S. Prusiner, unpublished experiments). Another possibility is that PrP^C exerts its influence on prion replication through some other proteins. Recent studies using ligand blots with [^{125}I]-PrP 27-30 have identified several putative PrP ligands or "Plis" of M_r from 45, 66 and 110 kDa (82). All of the Plis are acidic proteins with pIs between 4.5 and 5.0. The 45-kDa Pli (Pli 45) increases during scrapie infection. While Pli 45 is found exclusively in brain, other Plis of higher M_r were found in systemic tissues. Whether the Plis detected by ligand blots participate in either the conversion of a PrP precursor into PrP^{Sc} or the initiation of scrapie infection remains to be established. If the Plis identified by PrP 27-30 binding also bind PrP^C, then characterization of these molecules may offer some insight into the cellular function of PrP^C.

MULTIPLE FORMS OF PRIONS

Numerous studies have emphasized the association of scrapie infectivity with membranes (83, 84). Detergent extraction of membrane-containing fractions led to the production of infectious particles with a variety of sizes (85). It is now known that prions may exist in multiple forms, all containing high levels of infectivity (Figure 2). Interconversion of the multiple forms of prions depends upon the relative levels of detergent and lipid (11). Rods measuring 10 to 20 nm in diameter and 100 to 200 nm in length were found in purified preparations of scrapie prions (7, 9, 10, 18). Although previously we reported that the rods were generated by detergent extraction of microsomal membranes from scrapie-infected brains (21), we now know that both detergent extraction and limited proteolysis are required (86). Dispersion of the rods into DLPC was accomplished by addition of nondenaturing detergent and phospholipid.

Removal of detergent from the DLPC resulted in the formation of closed liposomes and subsequent removal of the phospholipid recreated the rods with full retention of scrapie prion infectivity. Dispersion of microsomal membranes in DLPC directly bypassing rod formation has also been accomplished (11, 12).

The rods that we observed in an earlier study, presumably resulted from the aggregation of PrP 27-30 molecules which were produced by endogenous proteases (21, 86). While most of the PrP immunoreactive protein in these detergent-extracted microsomes migrated on Western blots with a M_r of 33-35 kDa, some PrP molecules of M_r 27-30 kDa were also detected. In reports on the N-terminal sequence of hamster and mouse PrP^{Sc}, other investigators stated that their purified preparations contained numerous rod-like structures (mislabeled scrapie-associated fibrils, see below) (37, 70). Presumably, the rods in these fractions were generated from partially hydrolyzed PrP^{Sc} molecules which were not sequenced. These investigators dissociated the rods by

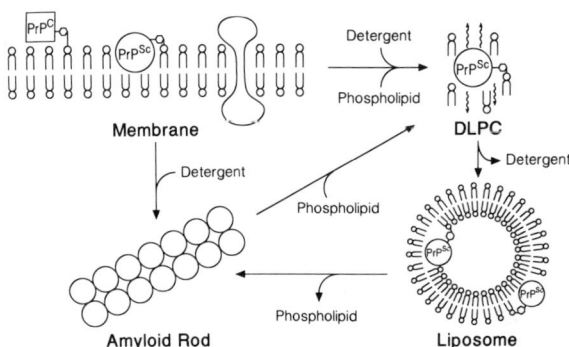

FIGURE 2. Interconversion of multiple prion forms. PrP^{Sc} is a membrane-bound protein which upon limited proteolysis produces PrP 27-30. Detergent extraction of membrane-bound PrP 27-30 produces amyloid rods which can be dispersed into DLPC by addition of phospholipids. DLPC can be formed directly from membranes by addition of detergent and phospholipid. Removal of detergents from DLPC produces closed liposomes. Removal of phospholipid from liposomes by organic solvent extraction (CHCl$_3$:methanol) regenerated the rods. All of the prion forms shown here possess high levels of scrapie infectivity (65).

denaturation in SDS prior to isolating PrP^{Sc} by gel filtration.

Sonication of the prion rods reduced their mean length to 60 nm and produced many spherical particles, without altering infectivity titers (20). The rods were found to be dissociated under nondenaturing conditions, with a combination of cholate and phosphatidylcholine (19). The resulting liposomes frequently showed a 10- to 50-fold increase in scrapie infectivity. Electron microscopy showed that the rods were completely disrupted upon formation of liposomes.

Although no unit morphologic structure could be identified for the rods, most have a relatively uniform diameter and often appear to be flattened cylinders. Some of the rods have a twisted structure, suggesting that they might be composed of protofilaments, but no consistent substructure could be discerned. Similar rod-shaped particles were isolated from brain tissue from patients dying of CJD (26). The heterogeneous morphology of the prion rods and the lack of consistent substructure distinguish them from viruses.

On the other hand, the ultrastructure of the prion rods is indistinguishable from that of many purified amyloids (10). Histochemical studies with Congo red dye have extended this analogy to purified preparations of prions (10), as well as to scrapie-infected brain tissue in which amyloid plaques have been shown to stain with antibodies to PrP 27-30 (87, 88). In addition, PrP 27-30 has been found to stain with periodic acid-Schiff reagent (6); amyloid plaques in tissue sections readily bind this reagent.

Immunocytochemical studies with antibodies to PrP 27-30 have shown that filaments (~16 nm in diameter and up to 1500 nm in length) within amyloid plaques of scrapie-infected hamster brain are composed of prion proteins (88). The prion filaments have a relatively uniform diameter, rarely show narrowings, and possess all the morphologic features of amyloids. Except for their length, the prion filaments appear to be identical ultrastructurally to the rods found in purified fractions of prions.

INFECTIOUS, SPORADIC AND GENETIC PRION DISEASES

The human prion diseases illustrate three mechanisms by which central nervous system degeneration may arise: slow infection, sporadic disease, and genetic disorder (89, 90). That the three diseases can be transmitted to laboratory animals by inoculation is well documented (91-93). Kuru is thought to have been spread exclusively through a slow infectious mechanism by means of ritualistic cannibalism (1, 94).

Although a few cases of CJD can be traced to inoculation with prions — i.e., injections of human growth hormone (30, 95, 96), transplantation of corneas, and implantation of cerebral electrodes — the vast majority appear to be sporadic, despite considerable effort to implicate scrapie-infected sheep as an exogenous source (1, 97). It is possible, although unlikely, that sporadic CJD results from prions that are ubiquitous in humans but have a very low efficiency of infection. In hamsters, scrapie infection by the oral route has been found to be 10^9 times less efficient than intracerebral inoculation (98).

Pedigree studies suggest that GSS and familial CJD may be inherited as autosomal dominant disorders (98-100). The significance of this observation was uncertain until recently when GSS was demonstrated to be a genetic disease and its development was linked to a missense variant of the prion protein (24, 25). Molecular cloning studies demonstrated a C to T substitution in the second position of codon 102 which probably results from deamination of a methylated C situated 5' to G (100, 102). This mutation creates a Dde 1 restriction site which was used to demonstrate genetic linkage between the PrP codon 102 amino acid substitution (Leu→Pro) and the development of GSS. Whether this substitution is the cause of GSS remains to be established. About 10 to 15 percent of cases of CJD are familial while most cases of GSS are inherited (90, 93, 99). GSS and familial CJD are the only known human diseases which are both genetic and infectious.

The genetic linkage of a missense variant of PrP with GSS constrains further the possible structural models for the prion. If prions contain a small, as yet undetected, nucleic acid, then this molecule must be widespread throughout the world to explain sporadic CJD and yet segregate in rare GSS families with the PrP mutation. It is noteworthy that Huntington's disease, another autosomal dominant genetic disorder, does not manifest clinical illness until the fifth decade in most cases. The delayed onset of both Huntington's disease and GSS is not understood. The genetic linkage results also suggest an interesting mechanism for sporadic CJD if prions contain only PrP^{Sc} (or PrP^{CJD}). A somatic mutation (103) in the PrP gene or even an RNA editing error (49-52) in a single cell might lead to the generation of PrP^{CJD} in that cell; prions would then spread to neighboring cells upon exit of the PrP^{CJD} molecules. In contrast, the PrP mutation of GSS is germline, consistent with its genetic transmission.

While scrapie is readily transmitted by inoculation and has been transmitted orally in an experimental setting (98, 104, 105), it is unclear how the disease spreads naturally amongst sheep and goats. Some investigators favored infection while others were proponents of a genetic mechanism (106, 107). In retrospect, both points of view were probably correct. Transmissible mink encephalopathy (TME) probably occurs in domesticated mink after the consumption of scrapie-infected sheep meat (2, 108). Scrapie prions administered intranasally to mink produce a disease indistinguishable from TME (109, 110). Interestingly, neither scrapie in mink nor TME can be transmitted to mice suggesting a significant species barrier for mink prions (111; W. Hadlow, unpublished data). Chronic wasting disease (CWD) in captive herds of mule deer and elk may have occurred after oral consumption of prion-contaminated sheep by-products (3). Whether CWD is a naturally occurring disease is uncertain. Over the past three years, over 1000 cases of dairy cattle in Great Britain have developed a spongiform degeneration (5, 112). Recent studies show that bovine spongiform degeneration (BSE) is transmissible to mice after a prolonged incubation period (113) and that a protease-resistant prion protein accumulates in the brains of these cattle (4). Epidemiological studies suggest that BSE began when bone meal fed to dairy cattle as a protein supplement was produced by a new process in which the rendering temperature was reduced around 1981 in order to increase the yield of sheep by-products (112). Whether BSE occurs naturally or it is due exclusively to the oral consumption of sheep scrapie prions is unknown. While the N-terminal sequence of PrP recovered from cattle brains is similar to that of hamster PrP (4), it will be important to learn about the PrP genes from a variety of cattle. Equally interesting will be the PrP sequences from many breeds of sheep as well as mink, mule deer and elk.

PRIONS ARE NOVEL AND UNPRECEDENTED

The convergence of experimental results indicating a pivotal role for PrP^{Sc} (or PrP^{CJD}) in animal and human prion diseases is both impressive and persuasive. The wide variety of independent disciplines including protein chemistry, molecular genetics, immunochemistry, neuropathology and experimental neurology, employed to obtain these experimental data, greatly strengthens the assertion that PrP^{Sc} has a central role in prion diseases.

That PrP is encoded by a cellular gene and not by a putative nucleic acid carried within the prion is a major feature which distinguishes prions from viruses. This discovery coupled with recent genetic studies showing linkage between a PrP missense variant and GSS demand that scrapie and CJD no longer be considered virological disorders. Although prion diseases resemble viral illnesses in some respects, the structure, cell biology and genetics of prions clearly separate them from viruses.

Whether prions are composed only of an abnormal isoform of the prion protein or they contain some additional molecule is uncertain. Many lines of evidence argue that PrP^{Sc} is the sole component of prions: (a) multiple forms of prions are infectious — membranes, rods, spheres, DLPC and liposomes; (b) many attempts to demonstrate the dependence of scrapie infectivity on a nucleic acid have been unsuccessful; (c) the ionizing radiation target size of the prion is 55,000 daltons; (d) PrP^{Sc} is encoded by a cellular gene; (e) mice with short and long incubation times have different PrP genes which encode distinct prion proteins and produce prions with distinct properties; and (f) GSS is linked to a mutation in the PrP gene. The partitioning of infectivity in a wide variety of different forms argues for a single component but does not eliminate the possibility of a second macromolecule. Numerous attempts to inactivate scrapie prion infectivity by procedures that hydrolyze or modify nucleic acids have been consistently unsuccessful. Ultraviolet irradiation of membranes, rods and DLPC suggests that if prions have an essential nucleic acid, then it will be <5 bases if single stranded or 30-45 bp if double stranded. Ionizing radiation studies give a target size too small to protect a large nucleic acid but do not rule out some other macromolecule.

Arguments in favor of a second prion component are: (a) prion infectivity has not been recovered from denatured samples after attempts at renaturation; and (b) many "strains" of prions have been reported. The first argument raises the possibility of a second component, but it need not necessarily be a nucleic acid. The second argument focuses on prion diversity and offers a nucleic acid genome as the basis for this diversity.

Just as GSS and familial CJD are unprecedented human illnesses since they are both genetic and infectious, the etiologic prion particles seem equally novel. Learning the chemical mechanisms responsible for converting PrP^C or a precursor into PrP^{Sc} will be extremely important. Elucidation of PrP^C function which may open up new avenues of research into mechanisms of cell homeostasis, recognition and possibly differentiation. Whether proteins other than PrP are converted from benign, cellular isoforms into malignant, diseased molecules resulting in degenerative diseases is unknown.

The search for other prion diseases should begin in earnest once the chemistry of PrP^{Sc} formation is elucidated. Whether some of the pathogenic mechanisms operative in prion diseases will give new insights into CNS disorders such as Alzheimer's disease, Parkinson's disease and amyotrophic lateral sclerosis, or some degenerative systemic diseases is unknown.

ACKNOWLEDGMENTS

Portions of this manuscript are adapted from an article to be published in Annual Review of Microbiology, Vol. 43, 1989. I am indebted to Drs. R. Barry, J. Bockman, D. Borchelt, D. Bredesen, D. Butler, S. DeArmond, R. Gabizon, K. Hsiao, S. Kent, V. Lingappa, D. Lowenstein, M. McKinley, B. Oesch, M. Rogers, M. Scott, D. Serban, N. Stahl, A. Taraboulos, E. Turk, and D. Westaway for important contributions, to Drs. G. Carlson, J. Cleaver, T. Crow, T. Diener, T. Endo, W. Hadlow, L. Hood, E. Kempner, D. Kingsbury, A. Kobata, D. Riesner, G. Roberts, J. Tateishi and C. Weissmann for collaborative studies that have been important to the progress of these investigations and are greatly appreciated, to D. Groth, L. Kenaga, C. Mirenda, A. Serban and M. Torchia for technical assistance, and to L. Gallagher for assistance in manuscript production.

REFERENCES

1. Gajdusek DC (1977). Unconventional viruses and the origin and disappearance of kuru. Science 197:943.
2. Hartsough GR, Burger D (1965). Encephalopathy of mink. I. Epizootiologic and clinical observations. J Infect Dis 115:387.

3. Williams ES, Young S (1980). Chronic wasting disease of captive mule deer: a spongiform encephalopathy. J Wildl Dis 16:89.
4. Hope J, Reekie LJD, Hunter N, Multhaup G, Beyreuther K, et al (1988). Fibrils from brains of cows with new cattle disease contain scrapie-associated protein. Nature 336:390.
5. Wells GAH, Scott AC, Johnson CT, Gunning RF, Hancock R. D, et al (1987). A novel progressive spongiform encephalopathy. Vet Rec 121:419.
6. Bolton DC, Meyer RK, Prusiner SB (1985). Scrapie PrP 27-30 is a sialoglycoprotein. J Virol 53:596.
7. Prusiner SB, Bolton DC, Groth DF, Bowman KA, Cochran SP, et al (1982). Further purification and characterization of scrapie prions. Biochemistry 21:6942.
8. Prusiner SB, Groth DF, Cochran SP, Masiarz FR, McKinley MP, et al (1980). Molecular properties, partial purification and assay by incubation period measurements of the hamster scrapie agent. Biochemistry 19:4883.
9. Prusiner SB, Groth DF, Bolton DC, Kent SB, Hood LE (1984). Purification and structural studies of a major scrapie prion protein. Cell 38:127.
10. Prusiner SB, McKinley MP, Bowman KA, Bolton DC, Bendheim PE, et al (1983). Scrapie prions aggregate to form amyloid-like birefringent rods. Cell 35:349.
11. Gabizon R, McKinley MP, Groth DF, Kenaga L, Prusiner SB (1988). Properties of scrapie prion liposomes. J Biol Chem 263:4950.
12. Gabizon R, McKinley MP, Groth DF, Prusiner SB (1988). Immunoaffinity purification and neutralization of scrapie prion infectivity. Proc Natl Acad Sci USA 85:6617.
13. McKinley MP, Bolton DC, Prusiner SB (1983). A protease-resistant protein is a structural component of the scrapie prion. Cell 35:57.
14. Barry RA, Prusiner SB (1986). Monoclonal antibodies to the cellular and scrapie prion proteins. J Infect Dis 154:518.
15. Oesch B, Westaway, D., Wälchli M, McKinley MP, Kent SBH, et al (1985). A cellular gene encodes scrapie PrP 27-30 protein. Cell 40:735.
16. Barry RA, Prusiner SB (1987). Immunology of prions. In Prusiner SB, McKinley MP (eds): "Prions — Novel Infectious Pathogens Causing Scrapie and Creutzfeldt-Jakob Disease," Orlando: Academic, p 239.

17. Ito M, Shinagawa M, Doi S, Sasaki S, Isomura H, et al (1988). Effects of the antiserum against a fraction enriched in scrapie-associated fibrils on the scrapie incubation period in mice. Microbiol Immunol 32:749.
18. Barry RA, McKinley MP, Bendheim PE, Lewis GK, DeArmond SJ, et al (1985). Antibodies to the scrapie protein decorate prion rods. J Immunol 135:603.
19. Gabizon R, McKinley MP, Prusiner SB (1987). Purified prion proteins and scrapie infectivity copartition into liposomes. Proc Natl Acad Sci USA 84:4017.
20. McKinley MP, Prusiner SB (1986). Biology and structure of scrapie prions. In Bradley RJ (ed): "International Review of Neurobiology," Vol 28, New York: Academic, p 1.
21. Meyer RK, McKinley MP, Bowman KA, Barry RA, Prusiner SB (1986). Separation and properties of cellular and scrapie prion proteins. Proc Natl Acad Sci USA 83:2310.
22. Carlson GA, Kingsbury DT, Goodman P, Coleman S, Marshall ST, et al (1986). Prion protein and scrapie incubation time genes are linked. Cell 46:503.
23. Westaway D, Goodman P, Mirenda C, McKinley MP, Carlson G, et al (1987). Distinct prion proteins in short and long scrapie incubation period mice. Cell 51:651.
24. Hsiao K, Baker HF, Crow TJ, Owen F, Poulter M, et al (1989). A prion protein missence variant is linked to Gerstmann-Sträussler syndrome. Clin Res, in press.
25. Hsiao KK, Westaway DA, Prusiner SB (1988). An amino acid substitution in the prion protein of ataxic Gerstmann-Sträussler syndrome. Am J Hum Genet 43:A87.
26. Bockman JM, Kingsbury DT, McKinley MP, Bendheim PE, Prusiner SB (1985). Creutzfeldt-Jakob disease proteins in human brains. N Engl J Med 312:73.
27. Bockman JM, Prusiner SB, Tateishi J, Kingsbury DT (1987). Immunoblotting of Creutzfeldt-Jakob disease prion proteins — host species-specific epitopes. Ann Neurol 21:589.
28. Bolton DC, McKinley MP, Prusiner SB (1982). Identification of a protein that purifies with the scrapie prion. Science 218:1309.
29. Bolton DC, McKinley MP, Prusiner SB (1984). Molecular characteristics of the major scrapie prion protein. Biochemistry 23:5898.
30. Gibbs CJ Jr, Joy A, Heffner R, Franko M, Miyazaki M, et al (1985). Clinical and pathological features and laboratory confirmation of Creutzfeldt-Jakob disease in

a recipient of pituitary-derived human growth hormone. N Engl J Med 313:734.
31. Kitamoto T, Tateishi J, Tashima T, Takeshita I, Barry RA, et al (1986). Amyloid plaques in Creutzfeldt-Jakob disease stain with prion protein antibodies. Ann Neurol 20:204.
32. Roberts GW, Lofthouse R, Brown R, Crow TJ, Barry RA, et al (1986). Prion protein immunoreactivity in human transmissible dementias. N Engl J Med 315:1231.
33. Butler DA, Scott MRD, Kingsbury DT, Bockman JM, Prusiner SB (1987). Murine neuroblastoma cell lines chronically infected with scrapie prions. Neurology 37 [Suppl]:342.
34. Kingsbury DT, Smeltzer D, Bockman J (1984). Purification and properties of the K. Fu. isolates of the agent of Creutzfeldt-Jakob disease. Sixth Intl Congr Virol, Sendai, Japan, p 70.
35. Race RE, Fadness LH, Chesebro B (1987). Characterization of scrapie infection in mouse neuroblastoma cells. J Gen Virol 68:1391.
36. Diringer H, Gelderblom H, Hilmert H, Ozel M, Edelbluth C, et al (1983). Scrapie infectivity, fibrils and low molecular weight protein. Nature 306:476.
37. Hope J, Morton LJD, Farquhar CF, Multhaup G, Beyreuther K, et al (1986). The major polypeptide of scrapie-associated fibrils (SAF) has the same size, charge distribution and N-terminal protein sequence as predicted for the normal brain protein (PrP). EMBO J 5:2591.
38. Manuelidis L, Sklaviadis T, Manuelidis EE (1987). Evidence suggesting that PrP is not the infectious agent in Creutzfeldt-Jakob disease. EMBO J 6:341.
39. Braig H, Diringer H (1985). Scrapie: concept of a virus-induced amyloidosis of the brain. EMBO J 4:2309-12
40. Czub M, Braig HR, Diringer H (1988). Replication of the scrapie agent in hamsters infected intracerebrally confirms the pathogenesis of an amyloid-inducing virosis. J Gen Virol 69:1753.
41. Chesebro B, Race R, Wehrly K, Nishio J, Bloom M, et al (1985). Identification of scrapie prion protein-specific mRNA in scrapie-infected and uninfected brain. Nature 315:331.
42. Czub M, Braig HR, Blode H, Diringer H (1986). The major protein of SAF is absent from spleen and thus not

an essential part of the scrapie agent. Arch Virol 91:383.
43. McKinley MP, Butler DA, Prusiner SB (1987). Prion protein mRNA in hamsters and mice. J Cell Biol 105:317a.
44. Rubenstein R, Kascsak RJ, Merz PA, Papini MC, Carp RI, et al (1986). Detection of scrapie associated fibril (SAF) proteins using anti-SAF antibody in non-purified tissue preparations. J Gen Virol 67:671.
45. Shinagawa M, Munekata E, Doi S, Takahashi K, Goto H, et al (1986). Immunoreactivity of a synthetic pentadecapeptide corresponding to the N-terminal region of the scrapie prion protein. J Gen Virol 67:1745.
46. Basler K, Oesch B, Scott M, Westaway D, Wälchli M, et al (1986). Scrapie and cellular PrP isoforms are encoded by the same chromosomal gene. Cell 46:417.
47. Locht C, Chesebro B, Race R, Keith JM (1986). Molecular cloning and complete sequence of prion protein cDNA from mouse brain infected with the scrapie agent. Proc Natl Acad Sci USA 83:6372.
48. Turk E, Teplow DB, Hood LE, Prusiner SB (1988). Purification and properties of the cellular and scrapie hamster prion proteins. Eur J Biochem 176:21.
49. Chen S-H, Habib G, Yang C-Y, Gu Z-W, Lee BR, et al (1987). Apoliportein B-48 is the product of a messenger RNA with an organ-specific in-frame stop codon. Science 238:363.
50. Feagin JE, Abraham JM, Stuart K (1988). Extensive editing of the cytochrome C oxidase transcript in Trypanosoma brucei. Cell 53:413.
51. Powell LM, Wallis SC, Pease RJ, Edwards YH, Knott TJ, et al (1987). A novel form of tissue-specific RNA processing produces apolipoprotein B-48 in intestine. Cell 50:831.
52. Shaw JM, Feagin JE, Stuart K, Simpson L (1988). Editing of kinetoplastid mitochondrial mRNAs by uridine addition and deletion generates conserved amino acid sequences and AUG initiation codons. Cell 53:401.
53. Sparkes RS, Simon M, Cohn VH, Fournier REK, Lem J, et al (1986). Assignment of the human and mouse prion protein genes to homologous chromosomes. Proc Natl Acad Sci USA 83:7358.
54. Kretzschmar HA, Prusiner SB, Stowring LE, DeArmond SJ (1986). Scrapie prion proteins are synthesized in neurons. Am J Pathol 122:1.

55. Lowenstein DH, Butler D, McKinley MP, Prusiner SB (1988). Chinese and Armenian hamster prion protein genes. J Cell Biol 107:136a.
56. Liao Y-C, Tokes Z, Lim E, Lackey A, Woo CH, Button JD, Clawson GA (1987). Cloning of rat "prion-related protein" cDNA. Lab Invest 57:370.
57. Goldman W, Hunter N, Multhaup G, Salbalm JM, Foster JD, et al (1988). The PrP gene in natural scrapie. Alzheimer's Disease and Associated Disorders — An International Journal 2 [Suppl]:331.
58. Westaway D, Prusiner SB (1986). Conservation of the cellular gene encoding the scrapie prion protein. Nucleic Acids Res 14:2035.
59. Barry RA, Kent SB, McKinley MP, Meyer RK, DeArmond SJ, et al (1986). Scrapie and cellular prion proteins share polypeptide epitopes. J Infect Dis 153:848.
60. Prusiner SB, McKinley MP, Groth DF, Bowman KA, Mock NI, et al (1981). Scrapie agent contains a hydrophobic protein. Proc Natl Acad Sci USA 78:6675.
61. Brown P, Coker-Vann M, Pomeroy K, Franko M, Asher DM, et al (1986). Diagnosis of Creutzfeldt-Jakob disease by Western blot identification of marker protein in human brain tissue. N Engl J Med 314:547.
62. Haraguchi T, Groth D, Barry RA, Fisher SJ, Teplow DB, et al (1987). Deglycosylation demonstrates two forms of the scrapie prion protein. Fed Proc 46:1319.
63. Robakis NK, Sawh PR, Wolfe GC, Rubenstein R, Carp RI, et al (1986). Isolation of a cDNA clone encoding the leader peptide of prion protein and expression of the homologous gene in various tissues. Proc Natl Acad Sci USA 83:6377.
64. Hay B, Barry RA, Lieberburg I, Prusiner SB, Lingappa VR (1987). Biogenesis and transmembrane orientation of the cellular isoform of the scrapie prion protein. Mol Cell Biol 7:914.
65. Bolton DC, Bendheim PE, Marmorstein AD, Potempska A (1987). Isolation and structural studies of the intact scrapie agent protein. Arch Biochem Biophys 258:579.
66. Bazan JF, Fletterick RJ, McKinley MP, Prusiner SB (1987). Predicted secondary structure and membrane topology of the scrapie prion protein. Protein Engineering 1:125.
67. Stahl N, Borchelt DR, Hsiao KK, Prusiner SB (1987). Scrapie prion protein contains a phosphatidylinositol glycolipid. Cell 51:229.

68. Stahl N, McKinley MP, Prusiner SB (1988). Differential release of cellular and scrapie prion proteins from cells by phosphatidylinositol phospholipase C. FASEB J 2:A989.
69. Stahl N, Borchelt DR, Taraboulos AT, Prusiner SB (1989). Differential release of prion protein isoforms from cloned scrapie-infected mouse neuroblastoma cells. J Cell Biol 107:387a.
70. Hope J, Multhaup G, Reekie LJD, Kimberlin RH, Beyreuther K (1988). Molecular pathology of scrapie-associated fibril protein (PrP) in mouse brain affected by the ME7 strain of scrapie. Eur J Biochem 172:271.
71. Bredesen DE, Scott MRD, Torchia T, Prusiner SB (1988). Cellular differentiation modulates prion protein expression and targeting. J Cell Biol 107:100a.
72. Giotta GJ, Heitzmann J, Cohn M (1980). Properties of two temperature-sensitive Rous sarcoma virus transformed cerebellar cell lines. Brain Res 202:445.
73. Prusiner SB (1982). Novel proteinaceous infectious particles cause scrapie. Science 216:136.
74. Bruce ME, Dickinson AG (1987). Biological evidence that scrapie agent has an independent genome. J Gen Virol 68:79.
75. Carlson GA, Goodman PA, Lovett M, Taylor BA, Marshall ST, Peterson-Torchia M, Westaway D, Prusiner SB (1988). Genetics and polymorphism of the mouse prion gene complex: the control of scrapie incubation time. Mol Cell Biol 8:5528.
76. Dickinson AG, Bruce ME, Outram GW, Kimberlin RH (1985). Scrapie strain differences: the implications of stability and mutation. In Tateishi J (ed): "Proceedings of Workshop on Slow Transmissible Diseases," Tokyo: Japanese Ministry of Health and Welfare, p 105.
77. Dickinson AG, Meikle VM (1971). Host-genotype and agent effects in scrapie incubation: change in allelic interaction with different strains of agent. Mol Gen Genet 112:73.
78. Dickinson AG, Outram GW (1979). The scrapie replication-site hypothesis and its implications for pathogenesis. In Prusiner SB, Hadlow WJ (eds): "Slow Transmissible Diseases of the Nervous System," Vol 2, New York: Academic, p 13.
79. Carlson GA, Westaway D, Prusiner SB (1988). The mouse prion gene complex and susceptibility to transmissible neurodegenerative diseases. Alzheimer's Disease and Associated Disorders 2:302.

80. McKinley MP, Mobley WC, Coleman S, Peterson M, Prusiner SB (1987). Developmental regulation of scrapie incubation times in neonatal Golden hamters. VIIth Intl Congr Virol, Edmonton, Alberta, Canada, p 147.
81. Mobley WC, Neve RL, Prusiner SB, McKinley MP (1988). Nerve growth factor increases mRNA levels for the prion protein and the beta-amyloid protein precursor in developing hamster brains. Proc Natl Acad Sci USA 85:9811.
82. Oesch B, Prusiner SB (1989). Identification of cellular proteins binding scrapie PrP. Clin Res, in press.
83. Hunter GD (1979). The enigma of the scrapie agent: biochemical approaches and the involvement of membranes and nucleic acids. In Prusiner SB, Hadlow WJ (eds): "Slow Transmissible Diseases of the Nervous System," Vol 2, New York: Academic, p 365.
84. Hunter GD, Millson GC (1967). Attempts to release the scrapie agent from tissue debris. J Comp Pathol 77:301.
85. Prusiner SB, Hadlow WJ, Garfin DE, Cochran SP, Baringer JR, et al (1978). Partial purification and evidence for multiple molecular forms of the scrapie agent. Biochemistry 17:4993.
86. McKinley MP, Meyer R, Kenaga L, Rahbar F, Serban A, et al (1988). Scrapie prion rod formation requires both detergent extraction and limited proteolysis. J Cell Biol 107:725a.
87. Bendheim PE, Barry RA, DeArmond SJ, Stites DP, Prusiner SB (1984). Antibodies to a scrapie prion protein. Nature 310:418.
88. DeArmond SJ, McKinley MP, Barry RA, Braunfeld MB, McColloch JR, et al (1985). Identification of prion amyloid filaments in scrapie-infected brain. Cell 41:221.
89. Alpers M (1987). Epidemiology and clinical aspects of kuru. In Prusiner SB, McKinley MP (eds): "Prions — Novel Infectious Pathogens Causing Scrapie and Creutzfeldt-Jakob Disease," Orlando: Academic, p 451.
90. Ridley RM, Baker HF, Crow TJ (1986). Transmissible and non-transmissible neurodegenerative disease: similarities in age of onset and genetics in relation to aetiology. Psychol Med 16:199.
91. Gajdusek DC, Gibbs CJ Jr, Alpers M (1966). Experimental transmission of a kuru-like syndrome to chimpanzees. Nature 209:794.

92. Gibbs CJ Jr, Gajdusek DC, Asher DM, Alpers MP, Beck E, et al (1968). Creutzfeldt-Jakob disease (spongiform encephalopathy): transmission to the chimpanzee. Science 161:388.
93. Masters CL, Gajdusek DC, Gibbs CJ Jr (1981). Creutzfeldt-Jakob disease virus isolations from the Gerstmann-Sträussler syndrome. Brain 104:559.
94. Alpers MP (1979). Epidemiology and ecology of kuru. In Prusiner SB, Hadlow WJ (eds): "Slow Transmissible Diseases of the Nervous System," Vol 1, New York: Academic, p 67.
95. Brown P (1988). The decline and fall of Creutzfeldt-Jakob disease associated with human growth hormone therapy. Neurology 38:1135.
96. Goujard J, Entat M, Maillard F, Mugnier E, Rappaport R, et al (1988). Human pituitary growth hormone (hGH) and Creutzfeldt-Jakob disease: results of an epidemiological survey in France, 1986. Int J Epidemiol 17:423.
97. Kovanen J, Haltia M (1988). Descriptive epidemiology of Creutzfeldt-Jakob disease in Finland. Acta Neurol Scand 77:474.
98. Prusiner SB, Cochran SP, Alpers MP (1985). Transmission of scrapie in hamsters. J Infect Dis 152:971.
99. Baker HF, Ridley RM, Crow TJ (1985). Experimental transmission of an autosomal dominant spongiform encephalopathy: Does the infectious agent originate in the human genome? Br Med J 291:299.
100. Masters CL, Gajdusek DC, Gibbs CJ Jr (1981). The familial occurrence of Creutzfeldt-Jakob disease and Alzheimer's disease. Brain 104:535.
101. Barker D, Schafer M, White R (1984). Restriction sites containing CpG show a higher frequency of polymorphism in human DNA. Cell 36:131.
102. Bird AP (1986). CpG-rich islands and the function of DNA methylation. Nature 321:209.
103. Hansen MF, Cavenee WK (1987). Genetics of cancer predisposition. Cancer Res 47:5518.
104. Pattison IH, Hoare MN, Jebbett JN, Watson WA (1972). Spread of scrapie to sheep and goats by oral dosing with foetal membranes from scrapie-affected sheep. Vet Rec 90:465.
105. Pattison IH, Millson GC (1961). Experimental transmission of scrapie to goats and sheep by the oral route. J Comp Pathol Therapeut 71:171.

106. Parry HB (ed) (1983). "Scrapie Disease in Sheep," New York: Academic.
107. Parry HB (1962). Scrapie: a transmissible and hereditary disease of sheep. Heredity 17:75.
108. Hadlow WJ, Karstad L (1968). Transmissible encephalopathy of mink in Ontario. Can Vet J 9:193.
109. Hadlow WJ, Race RE (1986). Cerebrocortical degeneration in goats inoculated with mink-passaged scrapie virus. Vet Pathol 23:543.
110. Hadlow WJ, Race RE, Kennedy RC (1987). Temporal distribution of transmissible mink encephalopathy virus in mink inoculated subcutaneously. J Virol 61:3235.
111. Marsh RF, Kimberlin RH (1975). Comparison of scrapie and transmissible mink encephalopathy in hamsters. II. Clinical signs, pathology and pathogenesis. J Infect Dis 131:104.
112. Wilesmith JW, Wells GAH, Cranwell MP, Ryan JBM (1988). Bovine spongiform encephalopathy: epidemiological studies. Vet Rec, in press.
113. Fraser H, McConnell I, Wells GAH, Dawson M (1988). Transmission of bovine spongiform encephalopathy to mice. Vet Rec 123:472.

SPERM RECEPTOR OLIGOSACCHARIDES AS MEDIATORS OF SPERM-EGG INTERACTIONS IN MICE

Jeffrey D. Bleil and Paul M. Wassarman

Department of Cell and Developmental Biology
Roche Institute of Molecular Biology
Roche Research Center
Nutley, New Jersey 07110

ABSTRACT During fertilization in mammals, sperm first bind to unfertilized eggs in a species-specific manner via sperm receptors located in the egg extracellular coat, or zona pellucida. The mouse sperm receptor is an 83,000 M_r zona pellucida glycoprotein, called ZP3, that contains both asparagine- (N-) and serine/threonine- (O-) linked oligosaccharides. A specific size-class of ZP3 O-linked oligosaccharides accounts for the glycoprotein's ability to function as a sperm receptor. Enzymatic removal or modification of certain sugars that constitute these oligosaccharides destroys the ability of ZP3 to function as a sperm receptor. These and other observations strongly suggest that carbohydrates play a fundamental role as mediators of sperm-egg interactions during fertilization in mammals.

INTRODUCTION

The interaction of sperm and egg during fertilization is perhaps the most frequently cited example of cellular recognition in plants and animals. In this context, it is generally held that certain macromolecules located on the surface of sperm and eggs interact with one another in a highly specific, complementary manner. It is these interactions that are thought to account for the relatively high degree of species-specificity that characterizes the fertilization process.

Among mammals, fusion of sperm and egg to form a zygote is preceded by species-specific binding of sperm to the egg extracellular coat, or zona pellucida (1-5). Such binding is mediated by "sperm receptors" located on the zona pellucida and "egg-binding proteins" located on the sperm head. Each motile sperm must be well anchored by its head to the zona pellucida before it can penetrate the extracellular coat and fuse with egg plasma membrane (i.e., fertilize the egg).

The mouse sperm receptor is an 83,000 M_r glycoprotein, called ZP3, that is present in more than a billion copies in the egg zona pellucida (3-9). ZP3 consists of a 44,000 M_r polypeptide chain (402 amino acids), 3 or 4 complex-type N-linked oligosaccharides, and an undetermined number of O-linked oligosaccharides (6,7,10-13). It is the oligosaccharides that are responsible for the glycoprotein's relatively low isoelectric point and extremely heterogeneous appearance on SDS-PAGE.

The ability of ZP3 to serve as sperm receptor is attributable to certain of its O-linked oligosaccharides, and not to its polypeptide chain (3-6,14,15). Removal of these oligosaccharides from purified ZP3 by mild alkaline hydrolysis (β-elimination) destroys the glycoprotein's sperm receptor activity *in vitro*. On the other hand, the O-linked oligosaccharides recovered following β-elimination of ZP3 (in the presence of $NaBH_4$) retain sperm receptor activity *in vitro*. The O-linked oligosaccharides that exhibit sperm receptor activity have an apparent average M_r of 3,900.

Which of the sugars that constitute ZP3 O-linked oligosaccharides are important for sperm receptor function? Recently, we addressed this question by enzymatically and chemically modifying purified ZP3 and ZP3-derived O-linked oligosaccharides, and then testing the modified substrates for sperm receptor activity *in vitro* (16,17). The results (discussed here) strongly suggest participation of a galactose, located in α-linkage at the nonreducing terminus of the O-linked oligosaccharides, in sperm receptor activity. Such findings may point to the involvement of modified lactosaminoglycans of ZP3 in mediating interactions between mouse sperm and eggs. In any case, overall, these findings provide additional support for the proposal that sugars serve as "recognition determinants" during fertilization in mammals.

RESULTS

In Vitro Sperm-Egg Binding ("Competition") Assay

When sperm are added to ovulated eggs and 2-cell embryos *in vitro*, within seconds the zona pellucida of eggs and embryos are covered with motile sperm. These sperm are loosely associated with the zona pellucida and can be removed easily by gentle pipetting with a broad-bore micropipet. This state of gamete adhesion is referred to as *attachment* and does not exhibit species-specificity. Shortly thereafter, contact between sperm and the egg zona pellucida becomes more

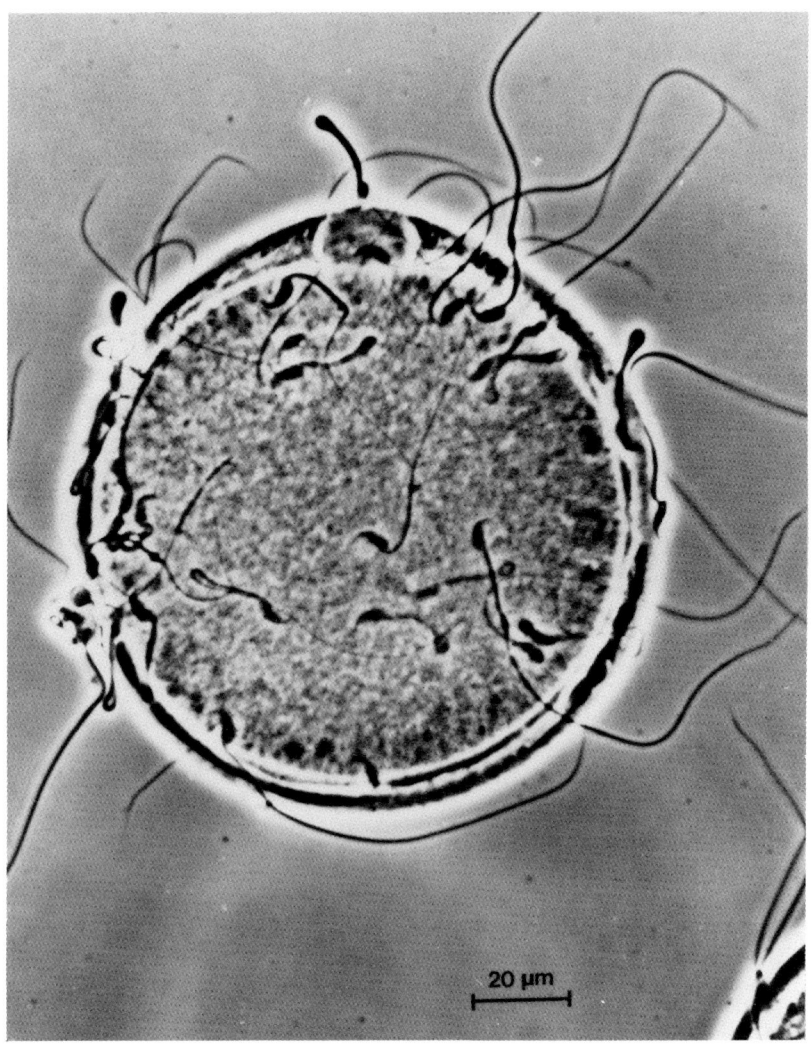

FIGURE 1. Light micrograph of sperm bound to an ovulated mouse egg *in vitro*. Eggs were incubated with sperm for 5 min, attached sperm removed, and eggs with bound sperm fixed in glutaraldehyde. Fixed eggs with bound sperm were transferred to a glass slide and gently "squashed" under a coverslip. The diameter of "squashed" eggs is about 1.3-times that of the untreated cell. The photomicrograph was taken by using a Zeiss 16X Plan-Neofluar objective.

tenacious, such that gentle pipetting no longer dissociates the gametes from one another. This state of gamete adhesion is referred to as *binding* and exhibits a certain degree of species-specificity (Fig. 1). It is bound sperm that are capable of penetrating the zona pellucida and fusing with egg plasma membrane.

Although the initial, reversible attachment of sperm to the embryo zona pellucida is virtually indistinguishable from that observed with eggs, in the former case attachment does *not* proceed to the binding state. This pronounced difference in behavior provides an operational definition of bound sperm as "those adhering to the egg zona pellucida under conditions that result in complete removal of sperm from the embryo zona pellucida".

This *in vitro* fertilization system provide a means of evaluating the potential "sperm receptor activity" of a given macromolecule. A *bona fide* sperm receptor should associate with the sperm head and thereby prevent binding of that sperm to the egg zona pellucida. The extent of inhibition of sperm binding, as compared to controls, reflects sperm receptor activity. This forms the basis of the "competition assay" used in experiments described here.

Purification of ZP3-Derived O-Linked Oligosaccharides

Purified ZP3 was prepared from mouse (CD-1; Charles River Laboratories) ovarian homogenates by gradient centrifugation (Percoll) and size fractionation (HPLC; TSK-250). Purified ZP3 was then subjected to β-elimination in the presence of NaB^3H_4, chromatographed on Dowex 50W-400, and size-fractionated by HPLC (Altex SW-2000). Fractions exhibiting sperm receptor activity (~3,900 M_r) were pooled and subjected to ion-exchange chromatography (HPLC; Altex DEAE-5PW). Active fractions were desalted and stored frozen prior to use.

Effect of Exoglycosidases on Sperm Receptor Activity

Purified ZP3 and ZP3-derived O-linked oligosaccharides inhibit binding of sperm to eggs *in vitro;* such inhibition of binding (i.e., *decrease* in number of sperm bound per egg, as compared to controls) is taken as an index of "sperm receptor activity". *Loss* of sperm receptor activity is reflected as an *increase* in the number of sperm bound per egg (i.e., approaching or equalling control values).

In order to assess involvement of individual sugars in sperm receptor activity, purified ZP3 and ZP3-derived O-linked oligosaccharides were treated with individual exoglycosidases. Sperm receptor activity of the substrates was determined by the competition assay, as described above, and release of sugars from the substrates (when sperm receptor activity was affected) was monitored by reverse-phase HPLC of dansylated sugar derivatives.

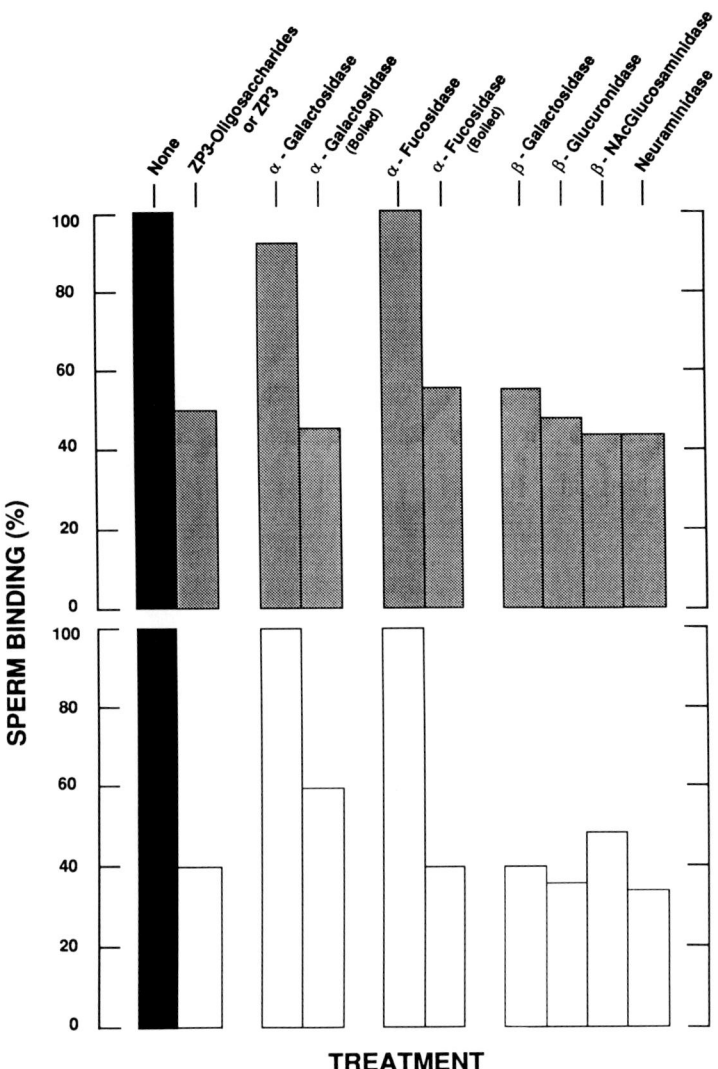

FIGURE 2. Effect of exoglycosidases on sperm receptor activity of ZP3 (white bars; bottom) and ZP3-derived O-linked oligosaccharides (stippled bars; top). The number of sperm bound to eggs in the presence of either ZP3 or ZP3-derived oligosaccharides was compared with the number bound in their absence (control; black bars) and expressed as a percentage of the control value.

The following commercially available exoglycosidases were tested in experiments described here: β-galactosidase (*E. coli*), α-galactosidase (coffee bean), β-glucuronidase (*E. coli*), α-fucosidase (beef kidney), β-N-acetylglucosaminidase (containing β-N-acetylgalactosaminidase; beef kidney), and neuraminidase (*C. perfringens*). Prior to use, each of the exoglycosidases was assayed under the appropriate conditions, using *o*- and *p*-nitrophenyl sugars as substrates, to determine the enzyme's specific activity.

Only α-galactosidase and α-fucosidase had a significant effect on the sperm receptor activity of ZP3 and ZP3-derived O-linked oligosaccharides, restoring sperm binding to control levels (Fig. 2). The other exoglycosidases did not have a significant effect on the sperm receptor activity of either substrate. Similarly, treatment of formaldehyde-fixed eggs with either α-galactosidase or α-fucosidase, but not with any of the other exoglycosidases, significantly reduced binding of sperm to the eggs. Sugar analyses revealed that galactose, and no other sugar, was released from ZP3-derived O-linked oligosaccharides by α-galactosidase. On the other hand, sugar analyses of α-fucosidase treated samples did not indicate removal of terminal fucose or any other sugar. Although we have no satisfactory explanation for the latter result, the former suggests that release of galactose, located in an α-linkage at the nonreducing terminus of ZP3 oligosaccharides, accounts for loss of sperm receptor activity.

Effect of Galactose Oxidase on Sperm Receptor Activity

To pursue the possible involvement of a terminal galactose in sperm receptor activity, purified ZP3 and ZP3-derived O-linked oligosaccharides were treated with galactose oxidase. This enzyme converts the C-6 position alcohol of terminal galactose and N-acetylgalactosamine residues in oligosaccharides to an aldehyde. A C-6 sugar alcohol can be regenerated on galactose oxidase-treated substrates by subsequent incubation in the presence of $NaBH_4$.

Consistent with the α-galactosidase results, treatment of either purified ZP3 or ZP3-derived O-linked oligosaccharides with galactose oxidase destroyed their sperm receptor activity (Fig. 3). Sugar analyses revealed that terminal galactose (not N-acetylgalactosamine) was, indeed, modified by galactose oxidase in these experiments. Furthermore, sperm receptor activity was restored to galactose oxidase-treated ZP3 and ZP3-derived O-linked oligosacchardies by subsequent reduction with $NaBH_4$ (Fig. 3). [It was noted that conversion of the sugar aldehyde to an oxime adduct with NH_2OH did not restore sperm receptor activity to galactose oxidase-treated ZP3-derived oligosaccharides.] Therefore, conversion of a ZP3 sugar alcohol to an aldehyde apparently is sufficient to prevent binding of sperm to ZP3.

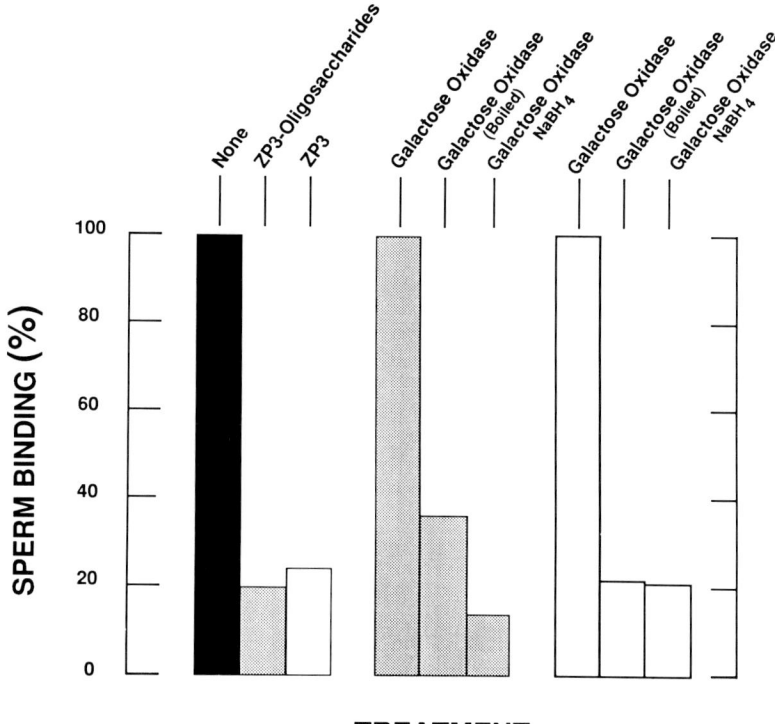

FIGURE 3. Effect of galactose oxidase on sperm receptor activity of ZP3 and ZP3-derived O-linked oligosaccharides. The number of sperm bound to eggs *in vitro* was determined under control conditions - i.e., in the absence of added ZP3 or ZP3-derived oligosaccharides (black bar), in the presence of untreated or galactose oxidase-treated (with or without $NaBH_4$) ZP3-derived oligosaccharides (sippled bars), or in the presence of untreated or galactose oxidase-treated (with or without $NaBH_4$) ZP3 (white bars). The number of sperm bound to eggs in the presence of either ZP3 or ZP3-derived oligosaccharides was compared with the number bound under control conditions and expressed as a percentage of the control value.

DISCUSSION

Results described here provide additional support for the idea that carbohydrates mediate species-specific binding of mammalian sperm to eggs This is an attractive idea in view of the great diversity of known oligosaccharide structures. The variety of compositions, sequences, branching patterns, conformations, and other features of oligosaccharides associated with glycoproteins provide for a staggering number of combinatorial possibilities. In fact, it is the structural diversity of oligosaccharides associated with glycoproteins that is utilized in their role as receptors for viruses, toxins, and hormones. Therefore, ZP3 is yet another case in which a glycoprotein's oligosaccharides are responsible for binding specificity.

Removal or modification of galactose, located in α-linkage at the nonreducing terminus of a specific size-class of ZP3 O-linked oligosaccharides, results in destruction of the glycoprotein's sperm receptor activity (Fig. 2). This suggests that galactose serves as at least one of the sperm receptor's "recognition determinants" during fertilization. Perhaps, most revealing is the finding that galactose oxidase-catalyzed modification of the terminal galactose is sufficient to destroy sperm receptor activity (Fig. 3). Thus, conversion of the sugar's C-6 alcohol to an aldehyde is sufficient to abolish binding of sperm to ZP3.

The results obtained with galactose oxidase/$NaBH_4$ are indicative of the highly specific nature of the molecular interactions that occur between receptor and sperm. The results are not unexpected, since interactions between carbohydrate binding proteins (in this case, "egg-binding proteins") and sugar ligands (in this case, ZP3 O-linked oligosaccharides) are stabilized largely by hydrogen bonds and van der Waals contacts, with the former providing the major contribution to binding (18). Sugar hydroxyls participate extensively in such hydrogen bonding, probably accounting for the inhibitory effects of galactose oxidase treatment on carbohydrate-mediated sperm-egg interaction.

Oligosaccharides with repeating N-acetyl-lactosamine units (i.e., lactosaminoglycans) are often associated with developmentally regulated glycoconjugates (19,20). In certain instances, such sequences are often terminated by a galactose in α-linkage with the galactose of the N-acetyl-lactosamine unit (20-22). In this context, preliminary results suggest that the carbohydrate epitope recognized by the monoclonal antibody designated LA4 (21,22), an epitope containing a terminal galactose in α-linkage with a penultimate galactose (structure based on the type 2 [galactose-β-1,4-N-acetyl-glucosamine] lactoseries), is present on ZP3 (R. Shalgi and P. Wassarman, unpublished results). Furthermore, it has been reported that lactosaminoglycans are present on a porcine zona pellucida glycoprotein that exhibits sperm receptor activity *in vitro* (23). It will be necessary to examine the potential role of ZP3 lactosaminoglycans in sperm receptor activity.

REFERENCES

1. Gwatkin, RBL (1977). "Fertilization Mechanisms in Man and Mammals." New York: Plenum Press.
2. Yanagimachi, R (1988). Mammalian fertilization. In Knobil, E, Neill, JD (eds): "The Physiology of Reproduction," vol 1, New York: Raven Press, p 135.
3. Wassarman, PM (1987). The biology and chemistry of fertilization. Science 235: 553.
4. Wassarman, PM (1987). Early events in mammalian fertilization. Annu Rev Cell Biol 3: 109.
5. Wassarman, PM (1988). Fertilization in mammals. Scientific American 255 (December): 78.
6. Wassarman, PM, Bleil, JD, Florman, HM, Greve, JM, Roller, RJ, Salzmann, GS, Samuels, FG (1985). The mouse egg's receptor for sperm: what is it and how does it work? Cold Spring Harbor Symp Quant Biol 50: 11.
7. Wassarman, PM (1988). Zona pellucida glycoproteins. Annu Rev Biochem 57: 415.
8. Bleil, JD, Wassarman, PM (1980). Mammalian sperm-egg interaction: identification of a glycoprotein in mouse egg zonae pellucidae possessing receptor activity for sperm. Cell 20: 873.
9. Bleil, JD, Wassarman, PM (1986). Autoradiographic visualization of the mouse egg's sperm receptor bound to sperm. J Cell Biol 102: 1363.
10. Salzmann, GS, Greve, JM, Roller, RJ, Wassarman, PM (1983). Biosynthesis of the sperm receptor during oogenesis in the mouse. EMBO J 2: 1451.
11. Roller, RJ, Wassarman, PM (1983). Role of asparagine-linked oligosaccharides in secretion of glycoproteins of the mouse egg's extracellular coat. J Biol Chem 258: 13243.
12. Kinloch, RA, Roller, RJ, Fimiani, CM, Wassarman, DA, Wassarman, PM (1988). Primary structure of the mouse sperm receptor's polypeptide chain determined by genomic cloning. Proc Natl Acad Sci, USA 85: 6409.
13. Ringuette, MJ, Chamberlin, ME, Baur, AW, Sobieski, DA, Dean, J (1988). Molecular analysis of cDNA coding for ZP3, a sperm binding protein of the mouse zona pellucida. Devl Biol 128: 78..
14. Florman, HM, Bechtol, KB, Wassarman, PM (1984). Enzymatic dissection of the function of the mouse egg's receptor for sperm. Devl Biol 106: 243.

15. Florman, HM, Wassarman, PM (1985). O-Linked oligosaccharides of mouse egg ZP3 account for its sperm receptor activity. Cell 41: 313.
16. Bleil, JD, Wassarman, PM (1988). Galactose at the nonreducing terminus of O-linked oligosaccharides of mouse egg zona pellucida glycoprotein ZP3 is essential for the glycoprotein's sperm receptor activity. Proc Natl Acad Sci, USA 85: 6778.
17. Wassarman, PM (1989). Carbohydrate-mediated sperm-egg interactions during mammalian fertilization. In Feizi, T, Ruoslahti, E (eds): "Carbohydrate Recognition in Cellular Function," Ciba Foundation Symposia (number 145), New York: John Wiley, in press.
18. Quiocho, FA (1986). Carbohydrate-binding proteins: tertiary structures and protein-sugar interactions. Annu Rev Biochem 55: 287.
19. Pink, JRL (1980). Changes in T-lymphocyte glycoprotein structures associated with differentiation. Contemp Topics Mol Immunol 9: 89.
20. Feizi, T, Kapadia, A, Gooi, HC, Evans, MJ (1981). Human monoclonal antibodies detect changes in expression and polarization of the Ii antigens during cell differentiation in early mouse embryos and teratocarcinomas. In Muramatsu, T, Ikawa, Y (eds): "Teratocarcinoma and Cell Surface," Amsterdam: North Holland, p 167.
21. Dodd, J, Jessell, TM (1985). Lactoseries carbohydrates specify subsets of dorsal root ganglion neurons projecting to the superficial dorsal horn of rat spinal cord. J Neuroscience 5: 3278.
22. Jessell, TM, Dodd, J (1985). Structure and expression of differentiation antigens on functional subclasses of primary sensory neurons. Phil Trans Roy Soc Lond, Biol 308: 271.
23. Yurewicz, EC, Sacco, AG, Subramanian, MG (1987). Structural characterization of the M_r=55,000 antigen (ZP3) of porcine oocyte zona pellucida. J Biol Chem 262:564.

A LYMPHOCYTE HOMING RECEPTOR IS A LECTIN, A MEMBER OF THE EMERGING LEC-CAM FAMILY[1]

S.D. Rosen*, J.S. Geoffroy*, L.A. Lasky[†], M.S. Singer*, S. Stachel[†], L.M. Stoolman[∞] D.D. True*, and T.A. Yednock*[2]

* Dept. Of Anatomy, University of California, San Francisco, CA 94143-0452, [†] Dept. of Cardiovascular Research, Genentech, Inc., S. San Francisco, CA 94080, [∞] Dept. of Pathology, University of Michigan, MI 48109

ABSTRACT The migration of lymphocytes from the blood into most secondary lymphoid organs is initiated by a highly specific adhesive interaction between lymphocytes and the endothelium of specialized blood vessels known as high endothelial venules (HEV). The propensity of lymphocytes to migrate to particular lymphoid organs is referred to as lymphocyte homing and the receptors that dictate interactions with particular HEV are designated as homing receptors. Evidence is reviewed that the peripheral lymph node homing receptor is a lectin-like receptor that is identical to the cell surface antigen (gp90^{MEL-14}) defined by the MEL-14 monoclonal antibody. Molecular cloning of gp90^{MEL-14} reveals an extracellular region consisting of three tandem domains, homologous respectively to calcium-dependent animal lectins, epidermal

---

[1] Supported by NIH Grant GM23547 to SDR.
[2] Present address, Dept of Microbiology and Immunology, University of Calif., San Francisco, CA 94143

growth factor, and complement binding proteins. Two other proteins (ELAM-1 and GMP-140), one directly implicated as a cell adhesion molecule and the other suspected of being one, have recently been shown to have the same domain organization as gp90^{MEL-14}. The acronym LEC-CAM has been proposed as a name for this family.

THE PROCESS OF LYMPHOCYTE RECIRCULATION

An immune response to a foreign substance entering the body is initially organized in the lymphoid organ that drains the antigen's portal of entry. For example, a substance that enters through the skin will usually accumulate in a local draining lymph node, whereas an antigen that gains entry into the gut will be sequestered within a gut-associated lymphoid aggregate such as a Peyer's patch. The strategic problem faced by the immune system is how to deploy the vast repertoire of immunocompetent lymphocytes to these distributed lymphoid organs such that responsive lymphocytes, which are typically rare for any particular antigen, will come into contact with each substance that enters any given organ. The solution is the movement of large numbers of lymphocytes from organ to organ with the lymphocytes alternatively traveling between the blood and lymph compartments. This process is known as lymphocyte recirculation (Figure 1) and is considered in detail in several recent reviews (1, 2, 3, 4). The extraordinary efficiency of this process relies upon the massive flux of lymphocytes involved. For example, in the rat, the rate of flow of lymphocytes from the the thoracic duct (Figure 1) to the blood is sufficient to replace the entire complement of blood lymphocytes 10-20 times per day (5, 6). To maintain a steady state, there must be an equal efflux of lymphocytes from the blood.

Anatomic and physiologic experiments carried out by Gowans and coworkers (5, 7) first elucidated the pathway by which blood-borne lymphocytes gain entry into secondary lymphoid organs such as lymph nodes. Extravasation of a lymphocyte is initiated by a highly specific adhesive interaction between the lymphocyte and specialized cuboidal or columnar endothelial cells of postcapillary venules found

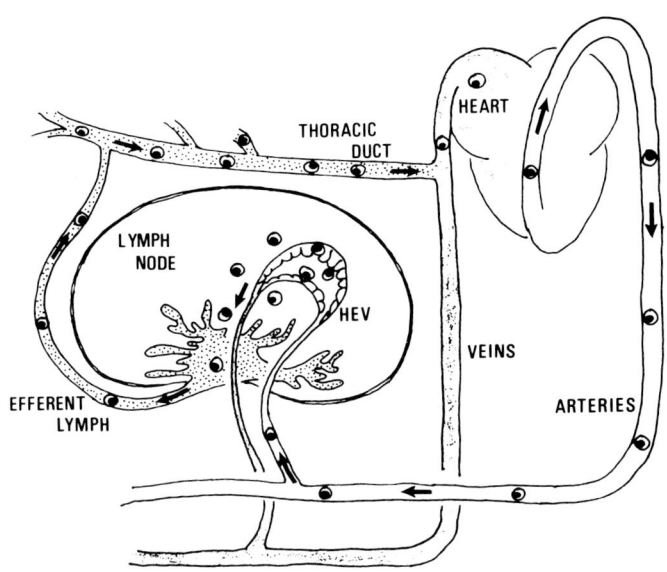

Fig. 1. Diagram of Lymphocyte Recirculation. Blood-borne lymphocytes arriving at a secondary lymphoid organ such as a lymph node encounter HEV in the cortex. A high percentage of cells adhere to the specialized endothelial cells and transmigrate across the endothelium. Cells that do not encounter cognate antigen will exit the node through the efferent lymphatics. The lymphocytes are returned to the blood through large lymphatic vessels such as the thoracic duct. Taken from (41) with permission.

within the lymphoid organ. Because of their distinctive morphology, these blood vessels are referred to as high endothelial venules or HEV. Attachment of lymphocytes occurs within seconds of contact followed by their migration between endothelial cells and through a basement membrane into the parenchyma of the lymphoid organ (8, 9).

Those lymphocytes that do not encounter cognate antigen in association with appropriate accessory cells (for antigen processing and presentation) will continue on the route of recirculation, leaving the lymphoid organ via lymphatics. Lymphocytes that become activated in response to antigenic stimulation will proliferate and differentiate within the lymphoid organ. Eventually, some of the differentiated progeny (e.g, effector and memory cells) will reenter the pathway of recirculation. HEV serve as the portal of entry into secondary lymphoid organs such as lymph nodes, Peyer's patches, tonsils, adenoids, appendix, and submucosal lymphoid aggregates in the respiratory, urogenital and GI tracts (10). Additionally, the entry of lymphocytes into certain sites of chronic inflammation appears to involve the transient induction of HEV-like blood vessels (11, 12). At some anatomic sites, lymphocytes are able to diapedese across thin-walled vessels (13).

THE REMARKABLE SPECIFICITY OF LYMPHOCYTE-ENDOTHELIAL ADHERENCE

It was clear from the earliest morphological studies that the interaction of lymphocytes with the endothelium of HEV is highly specific in that the prevalent leukocyte infiltrating the HEV wall is the lymphocyte (14). The application of the Stamper-Woodruff assay (Figure 2), a highly validated *in vitro* adherence assay (15), demonstrates that the interaction involves exquisitely refined adhesive specificities that depend on both the nature of the lymphocyte and the anatomical site of the HEV (reviewed in [2, 3, 4]). Parameters such as developmental stage, history of antigenic stimulation, lymphocyte class and subclass determine the ability of lymphocytes to bind to HEV in different lymphoid organs. Lymphocyte migration into peripheral lymph nodes (16, 17), Peyer's patches (18), and lung-associated lymphoid tissue (19, 20) appears to be controlled by three distinct adhesive specificities. Still another specificity may be involved in the migration of lymphocytes into the

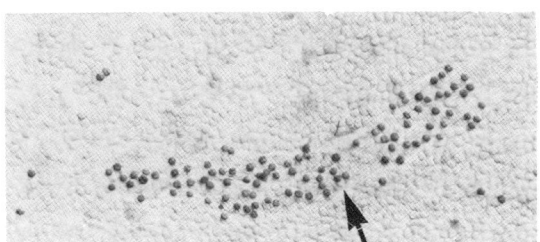

Fig. 2. Lymphocyte Attachment to HEV in the Stamper-Woodruff Assay. Cryostat-cut sections of lymphoid organs (e.g., lymph node) are overlaid with a viable suspension of lymphocytes. The lymphocytes attach very selectively to profiles of HEV exposed in the sections. The figure shows a longitudinal section through an HEV. The darker staining cells are exogenous lymphocytes that have attached during the assay. The arrow indicates the prominent basement membrane surrounding the HEV. From (41) with permission. Magnification ≈320X.

inflamed synovium associated with rheumatoid arthritis (21), although it is not yet clear whether this specificity is for joints or alternatively for sites of chronic inflammation. Figure 3 presents a model that accounts for selective lymphocyte migration into organized secondary lymphoid organs. It postulates the existence of a set of adhesive receptors (termed homing receptors) on lymphocytes and a complementary set of HEV-associated ligands which are recognized and bound by the receptors. Individual lymphocyte subpopulations may express these receptors in different combinations conferring upon them a repertoire of migratory (i.e., "homing") potentials, which may be very limited or quite broad. It has been proposed (2) that the migratory preferences of functional lymphocyte subpopulations dictated, at least in part, by the interaction of these homing receptor-HEV ligand pairs ensures that the local immune responses are both efficient and appropriate for the anatomic site.

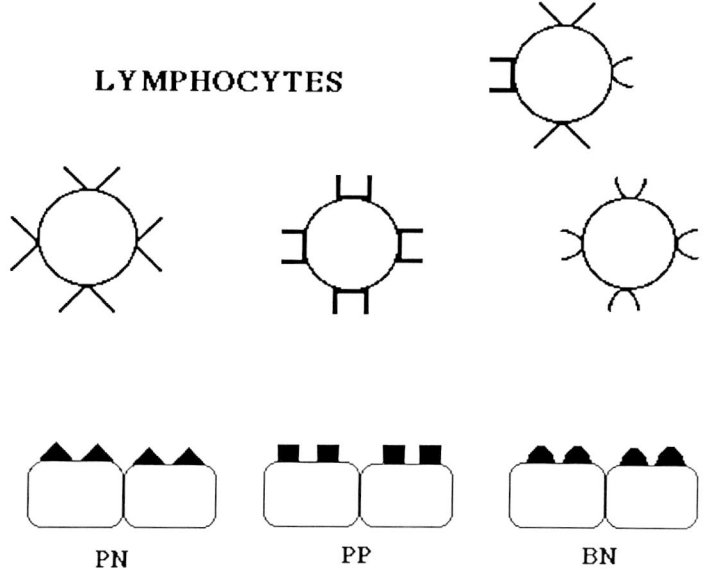

Fig. 3. A Model of Three Homing Specificities. The figure represents an extension of a model proposed by Butcher and Weissman (10) Three classes of homing receptors are shown, mediating lymphocyte adherence to HEV in peripheral lymph nodes (PN), Peyer's patches (PP), and bronchial lymph nodes (BN). Each type of HEV is proposed to display (on the lumenal aspect of its endothelial lining) a unique adhesive-ligand, which is complementary to one of the homing receptors. Lymphocytes may express a single homing receptor or a combination of all three homing receptors. Additional homing receptors may also exist. Although not depicted in this model, there is evidence that LFA-1, associated with lymphocytes, strengthens their interaction with HEV without imparting organ-selectivity (59, 60). A more detailed version of this model has been presented (4).

EVIDENCE FOR A CARBOHYDRATE-BINDING RECEPTOR ON LYMPHOCYTES

The enormous structural information that can potentially be encoded by carbohydrates makes these substances ideal candidates for recognition determinants in specific cell adhesion events (22). The receptors for these carbohydrate ligands are widely presumed to be proteins because of the demonstrated ability of lectins (23), antibodies (24, 25), and enzymes (26) to discriminate amongst an enormous variety of carbohydrate structures. Conclusive evidence for an involvement of carbohydrates in adhesive recognition derives from analyses of microbial-host cell interactions (27, 28). Less definitive but nonetheless very suggestive studies implicate specific carbohydrates in a wide range of eucaryotic cell-cell interactions (29, 30, 31, 32, 33, 34).

Early experiments aimed at evaluating the possible role of carbohydrates in lymphocyte adherence to HEV tested the ability of neutral monosaccharides to competitively inhibit the attachment of lymphocytes to rat lymph node HEV in the Stamper-Woodruff *in vitro* adherence assay (35). L-fucose and D-mannose selectively inhibit attachment at concentrations of ≥50 mM. Intriguingly, a relatively modest increase in the ionic strength of assay medium dramatically dramatically endhances carbohydrate-dependent inhibition by these neutral sugars, suggesting that charge is important in the interaction. When charged monosaccharides are tested as competitive inhibitors (36), mannose-6-phosphate (M6P) and the structurally related fructose-1 phosphate (F1P) are highly active relative to several other phosphorylated sugars. Phosphorylation of mannose at the 6-position augments its activity by 25-50 fold. In the rat system, 50% inhibition of lymphocyte binding to PN HEV is observed at 2-3 mM[3]. A further enhancement of inhibitory activity

---

[3] By way of analogy, it is interesting to note that the peptide, arg-gly-asp-ser (RGDS) is active in blocking various examples of cell-substratum interactions over a similar concentration range (61).

is observed with two polysaccharides, PPME and fucoidin. PPME is a mannan core derived from the yeast *Hansenula holstii*. It has a molecular weight of 2.5×10^6 daltons and consists exclusively of mannose and phosphate with one of every six sugar residues phosphorylated in the sixth position. PPME produces 50% inhibition of mouse lymphocyte attachment to PN HEV at 10-20 µg per ml which is equivalent to 4-8 nM of polysaccharide and 8-16 µM M6P. Fucoidin, containing a high concentration of fucose-4-sulfate but less well characterized than PPME, is more active on a mass basis than PPME. The effectiveness of these substances as inhibitors is not attributable to trivial physicochemical properties since a wide range of charged polysaccharides and glycosaminoglycans are largely inactive.

Preincubation experiments establish that both polysaccharides interact with the surface of lymphocytes to produce their inhibitory effects. A cytochemical probe, consisting of fluorescent beads covalently coupled to PPME, has allowed the lymphocyte surface receptor to be investigated in detail (37, 38). In correspondence with their inhibition of lymphocyte attachment to PN HEV, M6P and F1P selectively prevent PPME-bead attachment to lymphocytes. Several other parallels were established between PPME-bead binding to lymphocytes or lymphoma cell lines and the ability of these cells to attach to PN HEV. Among the most striking of these is the finding that selection of S49 lymphoma cells for increased ability to bind PPME-beads results in increased binding to PN HEV while not affecting their binding to other bead types or to PP HEV. Another important correspondence is that both interactions require calcium with identical quantitative dependencies on the extracellular concentration (Yednock, Singer, and Rosen, in preparation). Finally, Stoolman and Ebling (submitted) have recently shown that the phorbol ester, PMA, induces a correlated increase in PPME-binding and PN HEV attachment activities in a human lymphoblastic line (Jurkat). Taken together, these results suggest that the lymphocyte homing receptor involved in attachment to PN HEV is a calcium-dependent, lectin-like receptor. Evidence for a related receptor, in terms of carbohydrate-specificity, exists in mouse, rat and human (36, 37, 38, 39).

THE INVOLVEMENT OF CARBOHYDRATES ON HEV

The implication of a lectin-like receptor as a homing receptor has focused attention on the possibility that carbohydrate residues function as actual recognition determinants on HEV-ligands. The periodate sensitivity of the ligands associated with both PP and PN HEV as well as their resistance to protein denaturants are consistent with such a role (40, 41). More direct evidence derives from the demonstration that sialidases are able to selectively inactivate ligands on PN HEV without affecting those on PP HEV (40, 42). The same effects are seen whether the HEV were exposed to the enzymes *in vivo* or *in vitro* (42). Consistent with a selective inactivation of PN HEV ligands, infusion of sialidase into mice causes a marked inhibition of short-term lymphocyte accumulation within peripheral lymph nodes but not within Peyer's patches or a variety of nonlymphoid organs. The importance of sialic acid is further suggested by the finding that *Limax* agglutinin, a sialic acid-specific lectin, functionally inactivates HEV-ligands when it is incubated with tissue sections prior to the *in vitro* lymphocyte adherence assay (43). In contrast, a variety of other lectins encompassing a range of carbohydrate-binding specificities are inactive. Surprisingly, PP and PN HEV sites are equally susceptible to inhibition by *Limax* agglutinin. On the basis of these results, one can envision a role for sialic acid as either a recognition determinant of HEV-ligands or as a "modulator" of ligand activity (4, 42). The former possibility is very appealing in view of the existence of multiple forms and linkages of sialic acid in glycoconjugates[4] together with the known capacity of sialyloligosaccharides to encode highly specific ligands for both antibody recognition and microbial-host cell adhesive interactions. Regardless of the precise role of sialic acid in the function of HEV-ligands, the discovery that this sugar is essential for biological activity has interesting ramifications. For example, it is known that a variety of human pathogens cause markedly elevated levels of serum sialidase in infected hosts. It is tempting to speculate that a compromised immune system resulting from impaired lymphocyte recirculation may facilitate the spread of these pathogens.

RELATIONSHIP OF LECTIN-LIKE RECEPTOR TO $gp90^{MEL-14}$

Monoclonal antibodies against putative lymphocyte homing receptors have been described in the mouse (44, 45), rat (46, 47), and human systems (3, 48). The best characterized of these is MEL-14, which selectively blocks the interaction of mouse lymphocytes with PN HEV (44) without affecting attachment to PP HEV or to HEV in lung-associated lymph nodes. An excellent correlation has been established between the presence of $gp90^{MEL-14}$ and the ability of lymphocytes or lymphoma cells to bind to PN HEV. The antibody stains a majority of peripheral lymphocytes, and is present on other leukocytes including neutrophils, eosinophils, and monocytes but is absent on a variety of non lymphoid tissues (49). The antigen, as immunoprecipitated from surface radioiodinated cells, is a highly glycosyslated protein of molecular weight ≈90 kDa in lymphocytes (44) and slightly larger in neutrophils (49). Recently, a procedure has been devised (50) which allows the isolation of the antigen in high purity from detergent lysates of whole mouse spleens (Figure 4).

---

[4] The sialidase results with Peyer's patches may be explained by the fact that various sialyloligosaccharides are resistant to broad spectrum sialidases (64).

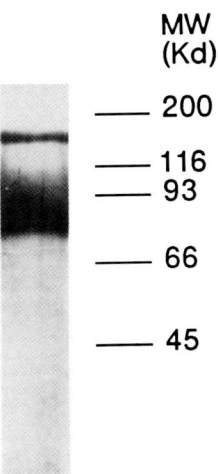

Fig. 4. SDS-PAGE Analysis of Immunoaffinity-Purified gp90^{MEL-14}. Antigen was purified from a detergent lysate of whole mouse spleen on a MEL-14 Sepharose column. The gel was silver stained. The broad band at 90 kDa corresponds in molecular weight to the antigen immunoprecipitated from surface radio-iodinated cells (44). N-terminal sequence derived from this material was used to design oligonucleotide probes for screening of cDNA libraries. The silver-stained band at 180 kDa is consistently observed in purified antigen preparations, although it represents $\leq$ 10% of the total mass. The relationship of this band to the 90 kDa component is not presently understood. Taken from (50) with permission.

A close relationship has been established between gp90^{MEL-14} on lymphocytes and the lectin-like receptor detected by PPME-beads (51). Firstly, PPME-bead binding is closely correlated with gp90^{MEL-14} expression in several lymphocyte and lymphoma populations. This correlation has recently been extended to lymphocytes that are exposed to proteases in the presence or absence of calcium (Yednock, Singer, and Rosen, in preparation). Secondly, MEL-14 antibody selectively blocks PPME attachment to lymphocytes, whereas a series of other cell surface-reactive antibodies are inactive. Moreover, MEL-14 does not affect lymphocyte binding of unconjugated beads or of beads derivitized with other charged polysaccharides (51). Finally, it has been shown in recent experiments (Geoffroy and Rosen, unpublished) that pre-exposure of PN tissue sections to isolated, soluble gp90^{MEL-14} prevents lymphocyte attachment to HEV[5]. If M6P is present during the pre-exposure step, the inhibition is reversed, indicating that the antigen is capable of a direct interaction with this sugar. This experiment has additional significance in that it provides the first direct evidence that gp90^{MEL-14} actually forms a bridge from the lymphocyte to the endothelium.

CONFIRMATION OF THE GP90^{MEL-14} AS A LECTIN

The results reviewed above provide very strong evidence that the gp90^{MEL-14} is a homing receptor for lymphocyte attachment to PN HEV and strongly argues that it functions as a lectin-like receptor. Molecular cloning of the antigen (50) confirms its classification as a lectin by showing that it belongs to a family of calcium-dependent (C-type) animal lectins (52). A full length cDNA was identified in a mouse spleen library with an oligonucleotide probe based upon the N-terminal sequence of isolated antigen. The largest open reading frame of the cDNA encodes a polypeptide of 372 amino acids. The sequence of the mature protein begins at residue 39, preceded by a sequence with the characteristics of a signal peptide. The next 294 residues are largely

---

[5] Chin, Woodruff and colleagues (62, 63) have identified a soluble factor, termed HEBF$_{PN}$, in rat lymph which has functional properties that are very similar to the isolated gp90^{MEL-14}. When PN sections are exposed to this factor, lymphocyte attachment to HEV is blocked. In its biochemical characteristics, there is, as yet, no clear relationship of this factor to gp90^{MEL-14}.

hydrophilic and contain 10 consensus sites for N-linked glycosylation, at least one of which is utilized. The next 22 residues are largely hydrophobic and likely constitute a transmembrane domain. Finally a short sequence of 17 amino acids at the C-terminus is presumably the cytosolic tail of the protein. Comparisons of the deduced sequence with the protein sequence databank reveals that the extracellular portion of the mature protein preceding the transmembrane domain can be divided into 3 distinct protein motifs. The first of these - from residues 39 to 155 - shows a highly significant degree of homology to carbohydrate recognition domains (CRDs) in several calcium-dependent animal lectins (Table 1). Members of this family of C-type lectins (52) exhibit a range of sugar-binding specificities that includes recognition of terminal galactose, N-acetylglucosamine, mannose, and fucose. Of the 18 amino acids identified by Drickamer (52) as being "almost invariant" in the CRDs of C-type lectins, 11 are conserved in the amino-terminal domain of the gp90^{MEL-14}. Following the lectin domain is a sequence of 34 residues with significant homology to epidermal growth factor (egf), a motif found in various cell surface receptors, developmental gene products, clotting factors etc. (53). The final discernable protein motif in the extracellular domain consists of two tandem and exact repeats of a 62 residue sequence. A highly homologous sequence is found in varying multiples in a number of proteins that exhibit binding to complement factors C3b or C4b (54). Other proteins containing this motif are not known to interact with complement factors. The overall domain structure of gp90^{MEL-14} is depicted in Figure 5.

The obvious conclusion from this molecular analysis is that the lectin domain of gp90^{MEL-14} is responsible for the calcium-dependent carbohydrate-binding activity of this protein, although formal proof will require the demonstration that the isolated CRD manifests this activity. It is particularly noteworthy that the putative CRD of MEL-14 possesses a very large number of lysine residues relative to the lectin domains of the previously identified C-type lectins. This feature may be pertinent since gp90^{MEL-14} interacts with anionic sugars whereas the other characterized C-type lectins bind to neutral sugars.

TABLE 1
COMPARISON OF AMINO-TERMINAL DOMAIN OF GP90^{MEL-14} TO CRD's OF C-TYPE ANIMAL LECTINS

Lectin	%Identity
IgE Receptor-Human#	29
Hepatic Lectin H2a-Human	28
Hepatic Lectin-Chicken	29
Acorn Barnacle Lectin	25
Mannose-Binding Protein C-Rat	26
Flesh Fly Lectin	23
Pulmonary Surfactant -Rabbit	21

#The IgE receptor has not as yet been shown to have lectin activity. Data taken from (51) with permission.

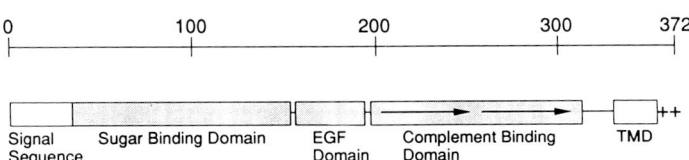

Fig. 5. Domain Structure of gp90^{MEL-14}. The figure summarizes the protein motifs found within the cloned gp90^{MEL-14}. These include a cleaved signal sequence, a C-type lectin domain, an epidermal growth factor (EGF) domain, two direct repeats of a complement-binding domain, a transmembrane domain (TMD) and a short, charged cytoplasmic tail at the C-terminus. Taken from (50) with permission.

The function of the other extracellular domains of the receptor are open to speculation. One can envision for either or both the purely structural role of placing the N-terminally-localized lectin domain in an appropriate orientation for ligand interaction or of allowing the molecule to establish functional interactions with other molecules in the lymphocyte membrane. The predicted occurrence of several disulfide bonds in these regions, based on comparisons with homologous regions of other proteins (55), lends support to this proposal. These domains could also be involved in stabilizing the interaction of the lymphocyte with the endothelium either during the initial cell-cell contact or as the lymphocyte progresses across the endothelium. A role in turnover of the receptor might be considered, particularly with respect to the "egf-like" domain. A further speculation is that the "complement binding domain" may be involved in preventing autologous attack of the lymphocyte cell surface by the complement cascade.

"HOMING" FOR THE FUTURE

A future focus will be the identification of the HEV-ligand for gp90^{MEL-14}. A prominent candidate is the structure recognized by MECA 79, a monoclonal antibody that reacts with the lumenal surface of PN HEV and selectively prevents lymphocyte attachment (56). Of critical interest is whether this antigen, or an associated molecule, contains a carbohydrate ligand complementary to the CRD of gp90^{MEL-14}. Should such a carbohydrate be identified, a detailed structural analysis may reconcile the apparently conflicting results of the sialidase/*Limax* agglutinin experiments and the sugar inhibition studies. We have previously speculated (4, 42) that the recognition determinant may be comprised of an unusual form of sialic acid, which is mimicked by M6P and F1P^6 .

---

[6] There are over 20 naturally occurring forms of sialic acid, which differ in the nature and position of O-linked and N-linked substituents (64).

A broader issue concerns the role of carbohydrates as recognition determinants in other lymphocyte homing specificities. As reviewed above, it is clear that lymphocytes can discriminate amongst HEV in different lymphoid organs. PPME does not block lymphocyte attachment to PP HEV, nor does sialidase treatment inactivate PP HEV ligands. Yet carbohydrates would appear to be involved in this interaction in view of the results of the periodate experiments and *Limax* agglutinin studies cited above. It remains to be determined whether a lectin of another carbohydrate specificity underlies the PP interaction. The speculation that there might be other lectin-like homing receptors gains credence from the recent discovery that two other molecules, ELAM-1 and GMP-140, bear striking resemblances to gp90^{MEL-14} in their molecular organization (55, 57). Each consists of an amino terminal C-type lectin domain (highly homologous to each other and to that in gp90^{MEL-14}), an "egf" domain, several tandemly arranged "complement-binding" domains, a putative transmembrane domain and a short C-terminal tail. ELAM-1 is biosynthetically induced on cultured endothelium by cytokines and mediates neutrophil attachment to the cytokine-treated endothelial cells. GMP-140 is present in platelets and endothelial cells in granules from which it is rapidly induced to the cell surface following cellular activation. The localization of GMP-140 as well as its structural homology to two known adhesion receptors suggests an adhesive function for this protein. By analogy to gp90^{MEL-14}, one strongly suspects that the amino-terminal lectin domains of ELAM-1 and GMP-140 will also prove to be integral to their adhesive functions. The unique extracellular domain structure of these three proteins as well as their suspected function as lectins suggests the acronym LEC-CAM (Lectin/ Egf/ Complement/ Cell/ Adhesion/ Molecule) to designate this family (58). It will be fascinating to determine whether the LEC-CAM family can be expanded beyond gp90^{MEL-14} to include lymphocyte homing receptors of other specificities.

REFERENCES

1. Ford WL (1975) Lymphocyte migration and immune responses. Prog Allergy 19: 1.
2. Butcher EC (1986). The regulation of lymphocyte traffic. Curr Top Micro Immunol 128: 85.
3. Woodruff JJ, Clarke LM, Chin YH (1987). Specific cell-adhesion mechanisms determining migration pathways of recirculating lymphocytes. Ann Rev Immunol 5:
4. Yednock TA, Rosen SD (1989) Lymphocyte homing. Adv Immunol 44: 313.
5. Gowans JL (1959) The recirculation of lymphocytes from blood to lymph in the rat. J Immunol 146: 54.
6. Ford WL (1969) The kinetics of lymphocyte recirculation through the rat spleen. Cell Tiss Kinet 2: 171.
7. Gowans JL, Knight EJ (1964). The route of recirculation of lymphocytes in the rat. Proc Roy Soc B 159: 257.
8. Bjerknes M, Cheng H, Ottaway CA (1986). Dynamics of lymphocyte-endothelial interactions *in vivo*. Science 232: 402.
9. Ottaway C. (1988). Dynamic aspects of lymphoid cell migration. In Husband, AJ (ed): "Migration and Homing of Lymphoid Cells," Florida: CRC Press, p 167.
10. Butcher EC, Weissman IL (1984). Lymphoid tissues and organs. In: Paul, WE: "Fundamental Immunology, " New York: Raven Press, p 109.
11. Smith JG, McIntosh GH, Morris B (1970). The migration of cells through chronically inflamed tissues. J Pathol 100: 21.
12. Freemont AJ (1983). A possible route for lymphocyte migration into diseased tissues. J Clin Pathol 36: 161.
13. Streeter PR, Berg EL, Rouse BTN, Bargatze RF, Butcher EC (1988). A tissue-specific

endothelial cell molecule involved in lymphocyte homing. Nature 331: 41.
14. Marchesi VT, Gowans JL (1964). The migration of lymphocytes through the endothelium of venules in lymph nodes: an electron microscopic study. Proc Roy Soc B 159: 282.
15. Stamper HB, Woodruff JJ (1976). Lymphocyte homing into lymph nodes: In vitro demonstration of the selective affinities of recirculating lymphocytes for high-endothelial venules. J Exp Med 144: 828.
16. Butcher EC, Scollay RG, Weissman IL (1980). Organ specificity of lymphocyte migration: mediation by highly selective lymphocyte interaction with organ-specific determinants on high endothelial venules. Eur J Immunol 10: 556.
17. Chin Y, Carey G, Woodruff J (1983). Lymphocyte recognition of lymph node high endothelium. V. Isolation of adhesion molecules from lysates of rat lymphocytes. J Immunol 131: 1368.
18. Chin YH, Rasmussen R, Cakiroglu AG, Woodruff JJ (1984). Lymphocyte recognition of lymph node high endothelium. VI. Evidence of distinct structures mediating binding to high endothelial cells of lymph nodes and Peyer's patches. J Immunol 133: 2961.
19. Geoffroy JS, Yednock TA, Curtis JL, Rosen SD (1988). Evidence for a distinct lymphocyte homing specificity involved in lymphocyte migration to lung-associated lymph nodes. FASEB J 2: A667.
20. Geoffroy JS, Yednock TA, Curtis JL, Rosen SD (1988). Further evidence for a lung-associated lymphocyte homing specificity. J Cell Biol 107: 552.
21. Jalkanen S, Steere AC, Fox RI, Butcher EC (1986). A distinct endothelial cell recognition system controlling lymphocyte traffic into inflamed synovium. Science 233: 556.
22. Sharon N. (1975). "Complex Carbohydrates. "Reading: Addison-Wesley
23. Lis H , Sharon N (1986). Lectins as molecules and as tools. Ann Rev Biochem 55: 35.

24. Feizi T (1985). Demonstration by monoclonal antibodies that carbohydrate structures of glycoproteins and glycolipids are onco-developmental antigens. Nature 314: 53.
25. Magnani JL (1987). Mouse and rat monoclonal antibodies directed against carbohydrates. Met Enzymol 138: 484.
26. Flowers H, Sharon N (1979). Glycosidases-properties and application to the study of complex carbohydrates and cell surfaces. Adv Enzymol. 48: 29.
27. Paulson JC (1985). Interaction of animal viruses with cell surface receptors. In Conn, PM (ed): "The Receptors, Vol. 2," New York: Academic Press.
28. Sharon N (1987). Bacterial lectins, cell-cell recognition and infectious disease. FEBS let 217: 145.
29. Grabel LB, Rosen SD, Martin G (1979). Teratocarcinoma stem cells have a cell surface carbohydrate-binding component implicated in cell-cell adhesion. Cell 17: 477.
30. Glabe CG, Grabel LB, Vacquier VD, Rosen SD (1982). Carbohydrate specificity of sea urchin bindin: a cell surface lectin mediating sperm-egg adhesion. J Cell Biol 94: 123.
31. Lopez LC, Bayna E, Litoff D, Shaper N, Shaper J, Shur BD (1985). Receptor function of mouse sperm surface galatosyltransferase during fertilization. J Cell Biol 101: 1501.
32. Crocker PR, Gordon S. (1986) Properties and distribution of a lectin-like hemagglutinin differentially expressed by murine stromal tissue macrophages. J Exp Med 164: 1862.
33. Künemund V, Jungalwala F, Fischer G, Chou D, Keihauer G, Schachner M. (1988) The L2/HNK-1 carbohydrate of neural cell adhesion molecules is involved in cell interactions. J Cell Biol 106: 213.
34. Bleil JD, Wassarman PM. (1988) Galactose at the nonreducing terminus of O-linked oligosaccharides of mouse zona pellucida glycoprotein ZP3 is essential for the glycoproteins's sperm.receptor activity.
35. Stoolman LM, Rosen SD (1983) Possible role for cell surface carbohydrate-binding

molecules in lymphocyte recirculation. J Cell Biol 96: 722.
36. Stoolman LM, Tenforde TS , Rosen SD (1984). Phosphomannosyl receptors may participate in the adhesive interactions between lymphocytes and high endothelial venules. J Cell Biol 99: 1535.
37. Yednock TA, Stoolman LM, Rosen SD (1987). Phoshomannosyl-derivatized beads detect a receptor involved in lymphocyte homing. J Cell Biol 104: 713.
38. Stoolman LM, Yednock TA , Rosen SD (1987). Homing receptors on human and rodent lymphocytes- evidence for a conserved carbohydrate-binding specificity. Blood 70: 1842.
39. Brandley BK, Ross TS , Schnaar RL (1987). Multiple carbohydrate-binding receptors on lymphocytes revealed by adhesion to immobilized polysaccharides. J Cell Biol 105: 991.
40. Rosen SD, Singer MS, Yednock TA , Stoolman LM (1985). Involvement of sialic acid on endothelial cells in organ-specific lymphocyte recirculation. Science. 228: 1005.
41. Rosen SD , Yednock TA. (1986) Lymphocyte attachment to high endothelial venules during recirculation: a possible role for carbohydrates as recognition determinants. Mol Cell Biochem 72: 153.
42. Rosen SD, Chi S, True DD, Singer MS , Yednock TA (1989). Intravenously injected sialidase inactivates attachment sites for lymphocytes on high endothelial venules. J Immunol (in press).
43. True DD , Rosen SD (1986). Sialic acid specific lectin blocks lymphocyte attachment to lymph node high endothelial venules in vitro. J Cell Biol 103: 194.
44. Gallatin WM, Weissman IL , Butcher EC (1983). A cell surface molecule involved in organ-specific homing of lymphocytes. Nature 303: 30.
45. Holzmann B, McIntyre BW, Weissman IL (1989). Identification of a murine Peyer's patch-

specific lymphocyte homing receptor as an integrin molecule with an alpha chain homologous to human VLA-4. Cell. 56: 37.
46. Rasmussen RA, Chin YH, Woodruff JJ , Easton TG (1985). Lymphocyte recognition of lymph node high endothelium. VII. Cell surface proteins involved in adhesion defined by monoclonal anti-HEBF-LN (A.11) antibody. J Immunol 135: 19.
47. Chin YH, Rasmussen RA, Woodruff JJ, Easton TG (1986). A monoclonal anti-HEBFPP antibody with specificity for lymphocyte surface molecules mediating adhesion to Peyer's patch high endothelium of the rat. J Immunol 136: 2556.
48. Jalkanen S, Bargatze R, Herron L, Butcher EC (1986). A lymphoid cell surface glycoprotein involved in endothelial recognition and lymphocyte homing in man. Eur J Immunol 10: 1195.
49. Lewinsohn DM, Bargatze RF, Butcher EC (1987). Leukocyte-endothelial cell recognition: evidence of a common molecular mechanism shared by neutrophils, lymphocytes, and other leukocytes. J Immunol 138: 4313.
50. Lasky LA, Singer MS, Yednock TA, Dowbenko D, Fennie C, Rodriguez H, Nguyen T, Stachel S, Rosen SD (1989). Cloning of a homing receptor reveals a lectin domain. Cell (in press).
51. Yednock TA, Butcher EC, Stoolman LM, Rosen SD (1987). Receptors involved in lymphocyte homing: relationship between a carbohydrate-binding receptor and the MEL-14 antigen. J Cell Biol 104: 725.
52. Drickamer K (1988). Two distinct classes of carbohydrate-recognition domains in animal lectins. J Biol Chem 263: 9557.
53. Doolittle RF, Feng DF, Johnson MS, McClure MA (1986). Relationships of human protein sequences to those of other organisms. CSHS Symp 51: 447.
54. Reid KBM, Bentley DR, Campbell RD, Chung LP, Sim RB, Kristensen T, Tack BF (1986). Complement system proteins which interact with C3b or C4b. Immunol Today 7: 230.
55. Johnston GI, Cook RG, McEver RP (1989).

Cloning of GMP-140, a granule membrane protein of platelets and endothelium: sequence similarity to proteins involved in cell adhesion and inflammation. Cell (in press).
56. Streeter PR, Rouse BTN, Butcher EC (1988). Immunohistologic and functional characterization of a vascular addressin involved in lymphocyte homing into peripheral lymph nodes. J Cell Biol 107: 1853.
57. Bevilacqua MP, Spengeling S, Gimbrone MA Jr, Seed B (1989). Endothelial leukocyte adhesion molecule 1: an inducible receptor for neutrophils related to complement regulatory proteins and lectins. Science (In press).
58. Stoolman LM (1989). Adhesion molecules controlling lymphocyte migration. Cell (in press).
59. Hamann A, Jablonski-Westrich D, Duijvestijn A, Butcher EC, Baisch H, Harder R, Thiele H (1988). Evidence for an accessory role of LFA-1 in lymphocyte-high endothelium interaction during homing. J Immunol 140: 693.
60. Pals ST, den Otter A, Miedema F, Kabel P, Keizer GD, Scheper RJ, Meijer CJLM (1988) Evidence that leukocyte function-associated antigen-1 is involved in recircualtion and homing of human lymphocytes via high endothelial venules. J Immunol 140: 1851.
61. Ruoslahti E, Pierschbacher M (1987). New perspectives in cell adhesion: RGD and integrins. Science. 238: 491.
62. Chin Y, Carey G, Woodruff J (1980). Lymphocyte recognition of lymph node high endothelium. II. Characterization of an in vitro inhibitory factor isolated by antibody affinity chromatography. J Immunol 125: 1770.
63. Chin Y, Carey G, Woodruff J (1982). Lymphocyte recognition of lymph node high endothelium. J Immunol 129: 1911.
64 Schauer R (1982). Chemistry, metabolism, and biological functions of sialic acids. Adv Carbo Chem Biochem 40: 131.

LYMPHOCYTE HOMING RECEPTORS AND VASCULAR ADDRESSINS[1]

Ellen Lakey Berg, Leslie A. Goldstein, Louis J. Picker, Philip R. Streeter[2], Mark A. Jutila, Robert F. Bargatze, David F. H. Zhou, and Eugene C. Butcher

Department of Pathology, Stanford University, Stanford, California 94305 and the Veterans Administration Medical Center, Palo Alto, California 94304.

ABSTRACT The exit of lymphocytes from the blood and entrance into lymphoid tissues is initiated by their adhesion to endothelial cells lining specialized post capillary venules or HEV. Lymphocytes employ tissue specific homing receptors to distinguish between HEV in different tissues. The tissue specific HEV determinants recognized by lymphocytes, the vascular addressins, have been identified in peripheral lymph nodes and mucosal lymphoid organs. The mucosal addressin has been demonstrated to specifically interact with a human lymphocyte homing receptor, $gp90^{Hermes}$. Complementary cDNAs for this homing receptor have been recently identified and indicate that $gp90^{Hermes}$

[1]This work was supported by NIH grants AI-19957 and GM-37734. ELB and MAJ are postdoctoral fellows of the American Cancer Society (National and California Divisions, respectively). LJP is a Career Development Awardee of the Veterans Administration. ECB is an Established Investigator of the American Heart Association.
[2]Present address: Systemix, Inc., Palo Alto, California.

shares homology to cartilage proteoglycan and link protein families.

INTRODUCTION

Lymphocytes continuously recirculate via the bloodstream and lymphatics, passing from blood to lymph selectively in lymphoid tissues. Recirculation of lymphocytes allows the full repertoire of antigenic specificities to be represented continuously throughout the body and probably facilitates specific cell-cell interactions which occur in distinct lymph node compartments and are required by various components of immune responses. The first step in this process of extravasation is the adhesion of lymphocytes to morphologically distinct, high endothelial venules (HEV) (1), found in lymphoid tissues and sites of chronic inflammation.

The most striking feature of lymphocyte-HEV interactions is that they are tissue specific (2). Employing short-term *in vivo* homing experiments, and an *in vitro* assay, originally developed by Stamper and Woodruff (3), in which lymphocytes bind specifically to HEV in frozen sections of lymphoid organs, at least three distinct lymphocyte-HEV recognition systems have been characterized. These include one mediating lymphocyte traffic to peripheral lymph nodes (PLN), another directing traffic to mucosal lymphoid organs such as the Peyer's patches (PP) or appendix, and a third controlling traffic to the inflamed synovium of arthritic joints (2,4). Recognition of synovial venules may be tissue-specific, or may involve a recognition system shared with other inflammatory tissues. Additional specificities have been more recently identified and include pulmonary lymphoid aggregates (5) and the skin (6). The existence of multiple tissue-specific lymphocyte-HEV recognition systems has been facilitated by the study of various murine and human lymphomas or transformed cell lines which bind selectively to either PLN or mucosal HEV (2,7). The subsequent generation of monoclonal antibodies which inhibit particular lymphocyte-HEV recognition events has confirmed the tissue-specificity of these interactions (see below) and allowed the identification of some of the participating molecules.

A second important feature of lymphocyte-HEV interactions is that HEV binding and the ability to recirculate is acquired and regulated during lymphocyte development and differentiation. Immature lymphocyte precursors, such as thymocytes, which do not recirculate, do not bind HEV whereas more developmentally mature, but virgin lymphocytes which recirculate into all lymphoid tissues bind well to the HEV in these tissues *in vitro*. More differentiated phenotypes representing antigen-specific effector and possibly memory lymphocyte populations migrate efficiently *in vivo* but often display more restricted HEV recognition and homing characteristics (reviewed in ref. 8).

Together these features suggest that a number of molecules under tight regulation are involved in the adhesion of lymphocytes to HEV. Quite recently, there have been significant advances in the identification and characterization of these molecules. Thus, in this paper we have chosen to summarize the studies from our laboratory, particularly those pertaining to the role of carbohydrate in lymphocyte-HEV interactions, and place them into context with the studies from other investigators.

LYMPHOCYTE HOMING RECEPTORS

Monoclonal Antibodies Define Homing Receptors

Monoclonal antibodies which inhibit lymphocyte-HEV binding have identified a number of lymphocyte cell surface molecules involved in the interaction with HEV (see Table 1). Those that are involved in tissue specific HEV recognition have been called homing receptors. In the mouse, MAb MEL-14 recognizes an 80-95 kD glycoprotein and blocks lymphocyte binding to peripheral node HEV (9). Antibodies of the Hermes series recognize

TABLE 1
LYMPHOCYTE ADHESION MOLECULES FOR HEV

Molecule	Mol. wt. (kD)	Gene family (or homology)	Antibody inhibition of lymphocyte binding to HEV[a]	Ref.
gp90^{Hermes} (CD44, Pgp-1)	85-95	proteoglycan/ link protein	PP (Hermes-3), PP, PLN, Syn (polyclonal anti-gp90^{Hermes})	4,10
MEL-14	85-95	mammalian lectin	PLN	9
HEBF$_{PP}$	80	?	PP	12
HEBF$_{LN}$	40,63,135	?	PLN	13
LPAM-1	130,160	integrin	PP	15
LFA-1	95,165	integrin	PLN,PP,?Syn	16,17
CT 4	30	?	PLN,PP,?Syn	14

[a]The ability of antibodies to block lymphocyte binding to peripheral lymph node (PLN), mucosal (PP), or synovial (Syn) HEV is indicated.

distinct epitopes on a human 85-95 kD glycoprotein(s), gp90^{Hermes} (10,11). MAb Hermes-3 inhibits lymphocyte binding to mucosal HEV but not to PLN or synovial HEV, and a polyclonal anti-gp90^{Hermes} blocks lymphocyte binding to all HEV classes, including synovial vessels (10). In the rat, Chin et al. (12) have described a monoclonal antibody which recognizes an 80 kD glycoprotein, HEBF$_{PP}$, and blocks lymphocyte binding to PP HEV. This same group has reported the involvement of HEBF$_{LN}$, a multi-chain antigen defined by MAb A.11, in rat lymphocyte binding to PLN HEV (13). Kraal and coworkers have reported that antibodies to a 30 Kd guinea pig lymphocyte antigen, CT 4, inhibit lymphocyte-HEV interactions, although not in a tissue specific fashion (14).

The relationship of the MEL-14 antigen and gp90^{Hermes} to one another has been clarified by the recent molecular cloning of both (see below). However, the relationship of these molecules to $HEBF_{PP}$ and $HEBF_{LN}$ and/or their potential association with each other or the other lymphocyte surface molecules which are involved in lymphocyte-HEV interactions remain to be established.

Two members of the integrin family of adhesion molecules play significant roles in lymphocyte-HEV binding. LPAM-1 is a mouse α/β heterodimer related to human VLA-4 which is involved in lymphocyte homing to mucosal but not peripheral lymphoid tissues (15). Another member of the integrin family, LFA-1, participates in lymphocyte-HEV interactions, although this is thought to be in an accessory rather than primary role (16,17).

Biochemical Characterization and Molecular Cloning of gp90^{Hermes}

We have recently reported the cloning of cDNA for gp90^{Hermes} expressed in a mucosal HEV-binding cell line (18). The DNA sequence predicts a membrane protein with a C-terminal cytoplasmic domain that may exist in two forms (18,19), a hydrophobic transmembrane domain (23 aa), and an N-terminal extracellular domain of 248 aa. The extracellular domain can be subdivided into a region proximal to the plasma membrane which constitutes the immunodominant region of the molecule, and a distal N-terminal region which contains 6 cysteines and five of the seven potential N-glycosylation sites (Figure 1). Biochemical studies of the gp90^{Hermes} molecule indicate that it is an acidic, sulfated, and extensively glycosylated molecule, containing both N- and O-linked sugars (20). Furthermore, a small fraction of gp90^{Hermes} molecules is modified by the addition of chondroitin sulfate, appearing on SDS-PAGE at 180-200 kD. The Hermes cDNA clone predicts four possible sites for chondroitin sulfate attachment, as well as multiple potential O-glycosylation sites within the proximal extracellular domain. No biochemical differences have been detected in the gp90^{Hermes} molecules

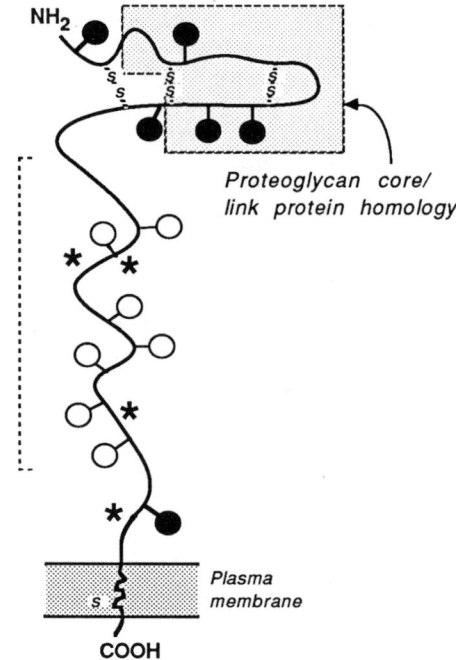

Figure 1. Proposed structure for gp90^{Hermes}. Potential attachment sites for N-linked sugars (closed ball and stick), O-linked sugars (open ball and stick) and chondroitin sulfate (*) are indicated. The region of homology with the cartilage proteoglycan core and link proteins is boxed and the region of immunodominance is indicated by the dashed line. Reprinted from Goldstein et al. (18), with permission.

isolated from mucosal versus peripheral lymph node HEV-binding lymphomas (20). In addition, Northern blots from these cell lines demonstrate identical mRNA species of 1.5, 2.2, and 4.5 kb (18). Thus, the role gp90^{Hermes} plays in conferring tissue specificity to lymphocyte-HEV interactions, if any, is not yet known.

Notably, a region within the distal N-terminal domain of gp90^{Hermes} displays striking homology to tandemly repeated domains in the cartilage proteoglycan core and link proteins (18,19). These repeated disulfide-linked loop domains are thought to be involved both in binding to hyaluronic acid and in link protein-proteoglycan core monomer interactions. The gp90^{Hermes} sequence is not related to the MEL-14 antigen or to the integrin, immunoglobulin or cadherin families of adhesion molecules. Recent cloning the MEL-14 antigen (S Rosen, this proceedings; and IL Weissman, personal communications) indicates that it displays a complex domain structure consisting of an N-terminal C-type lectin domain, an epidermal growth factor domain and consensus repeats similar to those found in complement regulatory proteins. Therefore, the MEL-14 antigen belongs to a new class of cell adhesion molecules which includes the cytokine inducible endothelial cell surface molecule ELAM-1 (see below), but which is distinct from gp90^{Hermes}.

Gp90^{Hermes}: a Novel Type of Cell Adhesion Molecule

Interestingly, antigens closely related to gp90^{Hermes} and recognized by the Hermes antibodies are expressed by a wide variety of non-lymphoid cells, including certain types of epithelial cells, fibroblasts, smooth muscle cells and a subset of glial cells (21). Biochemically, the gp90^{Hermes} antigens on glial cells and fibroblasts appear nearly identical to the 90 kD lymphocyte form. On squamous epithelium, however, Hermes antibodies predominantly recognize an antigen of 150 kD. This molecular species shares all Hermes antibody-defined epitopes with the other gp90^{Hermes} molecules, and is biochemically related to them. However, the epithelial cell Hermes antigen has a more basic pI and is insensitive to trypsin digestion. In addition, Stamenkovic et al. (19) have demonstrated that this form is derived from RNA species distinct from the lymphocyte form. Recently, gp90^{Hermes} has been shown to be identical to the CD44 and Pgp-1 antigens (11), and also the fibroblast extracellular matrix receptor III, which may be involved in fibroblast adhesion to extracellular matrix (22). Gp90^{Hermes} can interact with a variety of extracellular matrix components (S Jalkanen, personal communication). This suggests that gp90^{Hermes} may constitute a novel type of cell adhesion or recognition molecule potentially mediating cell-cell or cell-matrix interactions in multiple tissues.

VASCULAR ADDRESSINS

Monoclonal Antibodies Identify Recognition/ Adhesion Molecules for Lymphocytes on Endothelial Cells

The ability of lymphocytes to distinguish between HEV in different tissues implies that HEV express tissue-specific determinants for lymphocyte recognition. Monoclonal antibodies have allowed the identification of two of these vascular addressins, the mucosal (MAd) and peripheral lymph node (PNAd) addressins. MAb MECA-367, which recognizes MAd, selectively stains HEV in mucosal tissues and not in peripheral tissues, blocks lymphocytes from binding these HEV *in vitro* and also blocks the entry of lymphocytes into mucosal tissues *in vivo* (23). The PNAd specific MAb MECA-79 recognizes HEV selectively in peripheral lymph nodes and also blocks lymphocyte binding in a tissue-specific manner both *in vitro* and in *in vivo* short term homing studies (24).

These vascular addressins are distinct from other previously described endothelial cell adhesion molecules for leukocytes (Table 2). ELAM-1 is a 115 kD molecule whose expression is induced on cultures of human umbilical vein endothelium by IL-1 or TNF, and is thought to be involved in the extravasation of neutrophils into sites of acute inflammation in humans (25). The recent cloning of ELAM-1 indicates that it belongs to a novel group of cell surface proteins with a complex domain structure similar to that of the MEL-14 antigen (M Bevilacqua, personal communication). ICAM-1, a 97 kD member of the

TABLE 2
ENDOTHELIAL CELL ADHESION MOLECULES FOR LYMPHOCYTES
OR OTHER LEUKOCYTES

Molecule	Mol. wt. (kD)	Gene family or significant homology	Type of endothelium to which leukocytes are blocked from binding by antibodies[a]	Ref.
MAd	58-66	?	PP HEV	23
PNAd	92	?	PLN HEV	24
ELAM-1	115	mammalian lectin	activated HUVE	25
ICAM-1	97	immunoglobulin	activated HUVE	27

[a]Antibodies to MAd and PNAd block lymphocytes from binding to mucosal (PP) and peripheral lymph node (PLN) HEV, respectively. Antibodies to both ELAM-1 and ICAM-1 have been shown to block the adherence of neutrophils to cytokine activated cultures of human umbilical vein endothelium (HUVE).

immunoglobulin supergene family, is another cytokine-inducible endothelial cell molecule involved in leukocyte-endothelial cell interactions (26,27), which has been shown to be a ligand for the lymphocyte integrin molecule, LFA-1 (28). ICAM-1 has been found on a variety of cell types and participates in a wider variety of cell-cell interactions than the adhesion of leukocytes to endothelial cells (26,27). While these molecules may participate in the directed extravasation of leukocytes to inflammatory sites, they are not considered vascular addressins as their expression is not tissue specific.

Biochemical Features of the Vascular Addressins: Role of Carbohydrate in Lymphocyte-HEV Interactions

Carbohydrate containing structures on HEV have

been implicated in the adhesion of lymphocytes to HEV (29-31). The MEL-14 antigen, the homing receptor for peripheral lymph node HEV, has been demonstrated to be closely associated with a carbohydrate-binding activity on lymphocytes (31). The recent finding that the MEL-14 antigen displays homology with mammalian C-type lectins (see above) suggests that the MEL-14 antigen itself contains this activity. In preliminary studies, the putative HEV ligand for the MEL-14 antigen, PNAd, has been shown to differ quite substantially from its counterpart on mucosal HEV, MAd, in overall structure as well as in carbohydrate content (32). MAd is a single chain protein which migrates on gels at 54-62 kD under non-reducing conditions and at 58-66 kD under reducing conditions, suggesting the involvement of intra-chain disulfide bonds in its structure. On two dimensional gels, MAd displays charge heterogeneity with discrete pI's at 6.5, 6.6 and 6.7. The mucosal addressin appears to be resistant to N-glycanase, and binds to wheat germ agglutinin but not to lentil lectin. In contrast, the predominant species recognized by MECA-79 is a single chain glycoprotein of 92 kD, both under reducing and non-reducing conditions, and displays a single pI of 6.2. Furthermore, the peripheral addressin is sensitive to N-glycanase and binds to lentil lectin. Comparison of the sialic acid-containing structures on these two addressins will be of interest in light of the observation that lymphocyte interactions with peripheral lymph node HEV but not mucosal HEV have been shown to be sensitive to neuraminidase treatment (33).

The Mucosal Addressin is an Adhesion Molecule for Lymphocytes and a Ligand for gp90^{Hermes}

When incorporated into supported planar membranes, the mucosal addressin functions as an adhesion molecule for lymphocytes (34). Normal lymphocytes or mucosal HEV-binding lymphomas specifically adhere to planar membranes containing the mucosal addressin but not to membranes containing control proteins. This interaction is tissue selective in that neither PLN HEV-specific lymphomas nor HEV non-binding lymphomas adhere to the isolated MAd.

In addition, MAd appears to bind specifically to the lymphocyte Hermes homing receptor (35). Fluorescence energy transfer techniques were used to demonstrate that gp90^{Hermes} isolated from tonsil or from a mucosal HEV binding cell line, KCA, but not control glycophorin, displays saturable binding to MAd which can be specifically reversed with unlabeled Hermes antigen, but not with control proteins. This interaction is blocked by the monoclonal antibodies Hermes-3 and MECA-367 which inhibit lymphocyte binding to mucosal HEV. The results define a specific receptor-ligand (gp90^{Hermes}/MAd) interaction involved in lymphocyte adhesion to HEV. The ability of MAd to interact with other lymphocyte homing receptors/adhesion molecules remains to be tested.

ACKNOWLEDGEMENTS

The authors would like to acknowledge helpful discussions with Rupert Hallmann, Kei Kishimoto, Lusijah Rott and Steve Rosen.

REFERENCES

1. Gowans JL, Knight EJ (1964). The route of recirculation of lymphocytes in the rat. Proc. Roy. Soc. Ser. B 159:257.
2. Butcher EC, Scollay RG, Weissman IL (1979). Lymphocyte adherence to high endothelial venules: characterization of a modified *in vitro* assay and examination of the binding of syngeneic and allogeneic lymphocyte populations. Eur J Immunol 10:556.
3. Stamper HB, Woodruff JJ (1976). Lymphocyte homing into lymph nodes: *in vitro* demonstration of the selective affinity of recirculating lymphocytes for high endothelial venules. J Exp Med 144:828.
4. Jalkanen S, Steere A, Fox R, Butcher EC (1986). A distinct endothelial cell recog-

nition system that controls lymphocyte traffic into inflamed synovium. Science 233:556.
5. Geoffroy JS, Yednock TA, Curtis JL, Rosen SD (1988). Evidence for a distinct lymphocyte homing specificity involved in lymphocyte migration to lung associated lymph nodes. FASEB J 2:A667.
6. Chin Y-H, Falanga V, Streilein JW, Sachstein R (1988). Specific lymphocyte-endothelial cell interactions regulate migration into lymph nodes, Peyer's patches and skin. Reg Immunol 1:78.
7. Jalkanen S, Bargatze RF, Herron LR, Butcher EC (1986). A lymphoid cell surface glycoprotein involved in endothelial cell recognition and lymphocyte homing in man. Eur J Immunol 16:1195.
8. Butcher EC (1986). The regulation of lymphocyte traffic. Curr Top Microbiol Immunol 128:85.
9. Gallatin WM, Weissman IL, Butcher EC (1983). A cell surface molecule involved in organ-specific homing of lymphocytes. Nature 304:30.
10. Jalkanen S, Bargatze RF, de los Toyos J, Butcher EC (1987). Lymphocyte recognition of high endothelium: antibodies to distinct epitopes of an 85-95 kD glycoprotein antigen differentially inhibit lymphocyte binding to lymph node, mucosal or synovial endothelial cells. J Cell Biol 105:983.
11. Picker LJ, de los Toyos J, Telen MJ, Haynes BF, Butcher EC (1989). Identity of CD44, Pgp-1 and the Hermes class of lymphocyte homing receptors. J Immunol (in press).
12. Chin, Y-H, Rasmussen RA, Woodruff JJ, Easton TG (1986). A monoclonal anti-HEBF$_{pp}$ antibody with specificity for lymphocyte surface molecules mediating adhesion to Peyer's patch high endothelium of the rat. J Immunol 136:2556.
13. Rasmussen RA, Chin Y-H, Woodruff JJ, Easton TG (1985). Lymphocyte recognition of lymph node high endothelium. VII. Cell surface proteins involved in adhesion defined by monoclonal anti-HEBF$_{LN}$ (A.11) antibody. J

Immunol 135:19.
14. Kraal G, Twisk A, Tan B, Scheper R (1986). A surface molecule on guinea pig lymphocytes involved in adhesion and homing. Eur J Immunol 16:1515.
15. Holzmann B, McIntyre BW, Weissman IL (1989). Identification of a murine Peyer's patch-specific lymphocyte homing receptor as an integrin molecule with an α chain homologous to human VLA-4 α. Cell 56:37.
16. Hamman A, Jablonski-Westrich D, Duijvestijn A, Butcher EC, Baisch H, Harder R, Thiele H-G (1988) Evidence for an accessory role of LFA-1 in lymphocyte-high endothelium interaction during homing. J Immunol 140:693.
17. Pals ST, den Otter A, Miedema F, Kabel P, Deizer GD, Scheper RJ, Meijer CJ (1988). Evidence that leukocyte function associated antigen-1 is involved in recirculation and homing of human lymphocytes via high endothelial venules. J Immunol 140:1851.
18. Goldstein LA, Zhou DFH, Picker LJ, Minty CN, Bargatze RF, Ding JF, Butcher EC (1989). A human lymphocyte homing receptor, the Hermes antigen, is related to cartilage proteoglycan core and link proteins. Cell 56:1063.
19. Stamenkovic I, Amiot M, Pesando JM, Seed B (1989). A lymphocyte molecule implicated in lymph node homing is a member of the cartilage link protein family. Cell 56:1057.
20. Jalkanen S, Jalkanen M, Bargatze R, Tammi M, Butcher EC (1988). Biochemical properties of glycoproteins involved in lymphocyte recognition of high endothelial venules in man. J Immunol 141:1615.
21. Picker LJ, Nakache M, Butcher EC (1989). Monoclonal antibodies to human lymphocyte homing receptors define a novel class of adhesion molecules on diverse cell types. J Cell Biol. 109:927.
22. Gallatin W, Wayner E, St John T, Butcher EC, Carter W (1989). Structural homology between lymphocyte homing receptors and extracellular matrix receptor type III. (Submitted).
23. Streeter PR, Berg EL, Rouse BN, Bargatze RF,

Butcher EC (1988). A tissue-specific endothelial cell molecule involved in lymphocyte homing. Nature 331:41.
24. Streeter PR, Rouse BTN, Butcher EC (1988) Immunohistologic and functional characterization of a vascular addressin involved in lymphocyte homing into peripheral lymph nodes. J Cell Biol 107:1853.
25. Bevilacqua MP, Pober JS, Mendrick DL, Cotran RS, Gimbrone MA (1987). Identification of an inducible endothelial-leukocyte adhesion molecule. PNAS 84:9238.
26. Dustin ML, Rothlein R, Bhan AK, Dinarello CD, Springer TA (1986). Induction by IL 1 and interferon-gamma: tissue distribution, biochemistry and function of a natural adherence molecule (ICAM-1). J Immunol 137:245.
27. Smith CW, Rothlein R, Hughes B, Mariscalco M, Schmalstieg F, Anderson DC (1988). Identification of an endothelial determinant for CD18 dependent neutrophil adherence. FASEB J: A1237.
28. Marlin SD, Springer TA (1988). Purified intercellular adhesion molecule-1 (ICAM-1) is a ligand for lymphocyte function-associated antigen 1 (LFA-1). Cell 51:813.
29. Stoolman LM, Tenforde TS, Rosen SD (1984). Phosphomannosyl receptors may participate in the adhesive interactions between lymphocytes and high endothelial venules. J Cell Biol 96:1535.
30. Yednock TA, Butcher EC, Stoolman LM, Rosen SD (1987). Receptors involved in lymphocyte homing: relationship between a carbohydrate-binding receptor and the MEL-14 natigen. J Cell Biol 104:725.
31. Stoolman LM, Yednock TA, Rosen SD (1987). Homing receptors on human and rodent lymphocytes--evidence for a conserved carbohydrate binding specificity. Blood 70:1842.
32. Berg EL, Goldstein LA, Jutila MA, Nakache M, Picker LJ, Streeter PR, Wu NW, Zhou D, Butcher EC (1989). Homing receptors and vascular addressins: cell adhesion molecules that direct lymphocyte traffic. Imm Rev (in

press).
33. Rosen SC, Singer MS, Yednock TA, Stoolman LM (1985). Involvement of sialic acid on endothelial cells in organ-specific lymphocyte recirculation. Science 228:1005.
34. Nakache M, Berg EL, Streeter PR, Butcher EC (1989). The mucosal vascular addressin is a tissue-specific endothelial cell adhesion molecule for circulating lymphocytes. Nature 337:179.
35. Nakache M, Berg EL, Picker LJ, McConnell H, Butcher EC (1989). Specific binding of lymphocyte homing receptor to a tissue-specific vascular addressin. (Submitted).

THE RELATIONSHIP BETWEEN GOLGI AND CELL SURFACE β1,4 GALACTOSYLTRANSFERASE[1]

Adel Youakim[2] and Barry D. Shur

Department of Biochemistry and Molecular Biology
The University of Texas M.D. Anderson Cancer Center
Houston, Texas 77030

Abstract

β1,4 Galactosyltransferase (GalTase) is present in the Golgi apparatus and on the plasma membrane of most cells. However, the biochemical and structural relationship of the GalTases in these compartments is still unknown. In this article, we present arguments in support of the hypothesis that the cell surface form of GalTase is biochemically and molecularly distinct from the Golgi form of the enzyme.

Introduction

Over the past few years, it has become apparent that glycosyltransferases play a role in a wide variety of biological processes. The conventional view that glycosyltransferases are exclusively components of the Golgi apparatus and endoplasmic reticulum has been revised since these enzymes have also been localized to the plasma membrane. On the cell surface, glycosyltransferases function in intercellular adhesion, migration along the extracellular matrix, and possibly growth control. These findings support the hypothesis made some twenty years ago by Roseman (1970) and elaborated further by Roth and colleagues (Roth, 1973), in which cell surface glycosyltransferases were postulated to interact with substrate glycoconjugates on apposing cells or in the extracellular matrix. Since then, a variety of glycosyltransferases, such as fucosyltransferase (Rauvala et al., 1983), N-acetlygalactosaminyltransferase (Balsamo et al., 1986), sialyltransferase (Taatjes et al., 1988) and GalTase (for a review of cell surface

[1]This work was supported by grant HD22590 from the National Institutes of Health.
[2]Supported by a Fonds de la Recherche en Santé du Québec Fellowship.

GalTase see Shur, 1989) have been detected on cell surfaces. Of these, GalTase is the surface glycosyltransferase that has been studied most extensively. However, the relationship between the cell surface and the Golgi form of the enzyme is still poorly understood. Are they identical, similar or different? In this review, we will present arguments in support of the hypothesis that there exist at least two distinct forms of GalTase, one of which is localized to the Golgi apparatus and the other to the cell surface.

Properties of Golgi and Cell Surface GalTase

Depending on the source of material, GalTase is very diverse in size ranging in molecular weight from 45-80 kDa (for a review on GalTase see Strous, 1986). However, to date there has not been a comparative biochemical study of GalTase from the cell surface and Golgi. In spite of the paucity of available data, it has been suggested that GalTase at the cell surface is the net result of Golgi membrane turnover and fusion with the plasma membrane and thus, by implication, that the two enzymes are identical (Roth et al., 1985). Several immunohistochemical studies have demonstrated that GalTase is detectable at the cell surface using polyclonal antibody reagents that detect GalTase in the Golgi (Roth et al., 1985; Pestalozzi et al., 1982), suggesting that cell surface and Golgi GalTase share some common antigenic determinants. However, indirect evidence has accumulated to suggest that GalTase at the cell surface and in the Golgi are closely related, yet distinct proteins (see below).

In contrast to the biochemical data, enzymatic data on the two forms of GalTase are more abundant. A comparison of Golgi and cell surface GalTase reveals that they are enzymically indistinguishable. Their requirements for manganese, affinities for $\underline{N}$-acetylglucosamine (GlcNAc) and UDPgalactose (UDPGal), as well as their overall kinetic properties are similar (Cummings et al., 1979). Furthermore, the surface modifier protein α-lactalbumin alters the acceptor specificity of both enzymes from GlcNAc to glucose (Ebner and Magee, 1975).

The Golgi and cell surface forms of GalTase differ significantly in their functional roles. GalTase in the Golgi apparatus functions by catalyzing the transfer of Gal from UDPGal to terminal GlcNAc residues of oligosaccharides. In the case of cell surface GalTase, it is believed that the enzyme forms a stable complex with terminal GlcNAc residues on glycoproteins in the extracellular matrix or apposing cells, similar to a lectin or other carbohydrate binding protein. Since cells require a mechanism for dissociation from the extracellular matrix and from other cells, it is possible that GalTase operates catalytically or is shed from the cell surface to facilitate cell dissociation. However, such a catalytic mechanism would require the transport of sugar donors (e.g. nucleotide sugars) from the inside of the cell to the extracellular matrix, as was suggested by the studies of Turley and Roth (1979) in which cells spontaneously glycosylated the extracellular matrix coincident with migration upon it.

Evidence for Multiple Forms of GalTase Protein Structure

One obvious prerequisite in support of the hypothesis put forward in this review is that at least two GalTase proteins exist in cells possessing both surface and Golgi GalTase. In support of this, at least two proteins identified as GalTase have been purified from most tissues and cells examined. For example, in mouse F9 embryonal carcinoma cells (EC), GalTase has been identified as proteins of 59 kDa and 68 kDa (Suganuma et al., 1987), while in HeLa cells, proteins of 45 kDa, 47 kDa and 54 kDa have been reported to be GalTase (Strous et al., 1988). (For a more extensive list refer to Strous, 1986). However, it is unclear which of the proteins represents the Golgi and/or plasma membrane form of the enzyme. Furthermore, the structural relationship between the two forms of the enzyme is also unknown, but in HeLa cells the putative GalTase proteins were immunoprecipitated with the same antisera (Strous et al., 1988), suggesting some shared structural features. It is possible that different forms of GalTase proteins contain unique targeting sequences that route them to the appropriate cellular compartment analogous to those described for nuclear (Lanford et al., 1986) and endoplasmic reticulum proteins (Munro and Pelham, 1987). Alternatively, the different forms of GalTase may undergo posttranslational modifications responsible for subcellular routing similar to the Man-6-phosphate residues of lysosomal enzymes (Reitman and Kornfeld, 1981).

GalTase Molecular Biology

GalTase has been cloned from bovine (Narimatsu et al., 1986; Shaper et al., 1986), mouse (Shaper et al., 1988; Nakazawa et al., 1988; Lopez and Shur, 1988) and human (Appert et al., 1986; Masri et al., 1988) sources, all of which share >80% sequence homology.

The cloning data reveals the existence of multiple but highly related GalTase transcripts in some of the cells and tissues examined, which suggests these cells synthesize several similar but distinct proteins. For example, mouse mammary tumor cells synthesize two GalTase-specific mRNA transcripts, which are almost identical except that one contains a longer 5' end (Shaper et al., 1988). Using S1 nuclease protection assays, it was shown that the longer clone (4.1 kb) contained approximately 200 bp of non-coding and coding sequences not present in the shorter transcript (3.9 kb). The 4.1 kb transcript contains two potential translation start sites and encodes a protein of 399 amino acids, whereas the 3.9 kb transcript lacks the proximal translation start site as well as another 12 amino acids. These findings have led Shaper et al. (1988) to suggest a possible mechanism for differential membrane insertion of the proteins encoded by the two transcripts: the 4.1 kb transcript encodes a protein with one transmembrane domain and a putative cleavable N-terminal signal peptide, resulting in a cytoplasmic orientation of the C-terminal catalytic domain. The short transcript, on the other hand, which lacks the cleavable signal peptide would be oriented with its C-terminal in the lumen or extracellular enviroment. This model thus provides for a GalTase that may participate in cytoplasmic glycosylation in addition to the more traditional enzyme oriented lumenally/extracellularly. An alternative model provides for two GalTase proteins, both of which could be oriented lumenally/extracellularly: if the signal peptide of the long transcript is not cleaved, then the orientation of the encoded protein would be the same as the one specified by the shorter transcript (i.e. N-terminal cytoplasmic and C-terminal luminal/extracellular). Thus, the cell would possess two GalTases that are identical except for 13 amino acids at the N-terminal, which one can postulate, contains a sequence required for subcellular targeting (to the Golgi or the plasma membrane) or posttranslational modification. Evidence in support of an uncleaved signal peptide comes from the studies of Strous et al. (1988). They demonstrated that GalTase synthesized from HeLa cell mRNA translated in vitro was identical to GalTase made in intact cells in the presence of tunicamycin, thus precluding any proteolytic cleavage of the enzyme during translocation through the endoplasmic reticulum. In this same study, the existence of two GalTase mRNAs in HeLa cells was inferred since immunoprecipitation with anti-GalTase antisera of in vitro translated products revealed the presence of a 42 kDa and a 45 kDa protein (identical in size to GalTase made in tunicamycin-treated cells).

Narimatsu and colleagues have isolated several GalTase cDNAs from retinoic acid treated mouse F9 EC cells (Nakazawa et al., 1988). The major clone (λmGT-5) is identical to the long clone described by Shaper et al. (1988) and encodes a protein of 399 amino acides. In contrast to that found in mouse mammary cells, the other GalTase cDNA isolated from EC cells contains a 309 bp insertion in the coding sequence of λmGT-5 and a 35 amino acid deletion near the 3' end (Nakazawa et al., 1988). Thus the structural relationship between the two transcripts in mouse EC cells is different than that found in mouse mammary cells.

Differential Regulation of Cell Surface and Golgi GalTase

Perhaps the most provocative evidence in support of distinct Golgi and plasma membrane GalTase comes from the studies in which the activity of the enzyme in one compartment is altered without a concomitant affect on the activity in the other membrane compartment.

Rat parotid glands induced to hypertrophy by altered diet (Humphreys-Beher and Schneyer, 1987) or by isoproterenol treatment (Marchase et al., 1988) showed a four-fold (diet) to forty-fold (isoproterenol) increase in cell surface GalTase activity (Table 1). The activity of the Golgi-associated GalTase remained unchanged, thus suggesting that alterations in cell surface GalTase activity are not coupled to changes in Golgi activity. However, one can not exclude the possibility that the elevated cell surface GalTase activity results from increased synthesis and transport of GalTase from the Golgi to the cell surface (Marchase et al., 1988). This scenario would require that the flow of GalTase to the plasma membrane increases to such an extent that Golgi GalTase activity remains unchanged.

Recent studies of mesenchymal cell migration on various extracellular matrix components showed that cells migrating on laminin had three times the level of cell surface GalTase activity, as measured by enzyme activity and ^{125}I-antibody binding, than cells migrating on fibronectin (Table 1; Eckstein and Shur, 1989). However, total GalTase activity (cell surface and Golgi) was unchanged in cells migrating on either matrix, as was the level of GalTase mRNA. These results suggest that cell surface GalTase can increase without a concomitant increase in Golgi GalTase and support the notion that the two pools of GalTase are differentially regulated. However, the possibility remains that GalTase is stabilized on the cell surface due to association with laminin oligosaccharides, resulting in a slower turnover rate and increased accumulation.

Table 1

Experimental System	GalTase Activity		Reference
	Golgi	Cell surface	
Parotid gland hypertrophy[a]	NC[b]	++[c]	Marchase et al., 1988
			Humphreys-Beher & Scheyner, 1987
Migration on laminin[d]	NC	++	Eckstein & Shur, 1989
F9 EC cell differentiation[e]	++	NC	Lopez et al., 1989

[a] relative to normal parotid gland
[b] NC, no change
[c] ++, represents increased GalTase activity
[d] relative to fibronectin
[e] relative to undifferentiated cells

Differential regulation of Golgi and cell surface GalTase has also been demonstrated during EC cell differentiation. Mouse F9 EC cells treated with retinoic acid or retinoic acid-cAMP will differentiate in vitro into highly secretory endodermal cells with a well developed Golgi apparatus. In differentiated cells it was found that GalTase mRNA levels increased approximately three-fold compared to undifferentiated cells (Lopez et al., 1989). Subcellular fractionation of undifferentiated and differentiated cells indicated that cell surface GalTase specific activity remained relatively unchanged, whereas Golgi-associated GalTase specific activity increased five-fold (Table 1), a finding consistent with the accumulation of Golgi apparatus in differentiated cells. Therefore, in this instance, Golgi activity can be increased dramatically without a similar change in cell surface

GalTase, thus supporting the concept of independently regulated Golgi and cell surface pools of GalTase.

Conclusions and Future Perspectives

While the structural relationship between cell surface and Golgi GalTase is still unknown, evidence suggests that the enzymes are similar yet distinct. In many cells, two or more proteins have been identified as GalTase either by affinity chromatography or immunoprecipitation. Furthermore, immunolocalization with the same antibody reagent has detected GalTase in the Golgi and at the plasma membrane. In addition, molecular genetic data have revealed multiple GalTase transcripts that share high degrees of homology and, by extension, encode related proteins. Lastly, GalTase activity at the cell surface and in the Golgi can be differentially activated suggesting that the two pools of protein are independently regulated. Consequently, the multiplicity of GalTase proteins and mRNAs, coupled with the differential regulation of surface and Golgi GalTase specific activities, supports the notion that the two enzymes are biochemically and molecularly distinct.

The major issue that remains to be elucidated however, is the exact nature of the difference(s) between cell surface and Golgi GalTase. The availalility of GalTase cDNA will allow one to introduce genes for specific transcripts into cells and determine the proteins that they produce. In this way, one should be able to identify cell surface- and Golgi-specific genes and to determine which region(s) of the genes contain the information necessary for subcellular routing.

References

1. Roseman S (1970). The synthesis of complex carbohydrates by multiglycosyltransferase systems and their potential function in intercellular adhesion. Chem Phys Lipids 5:270.
2. Roth S (1973). A molecular model for cell interactions. Quart Rev Biol 48:541.
3. Rauvala H, Prieels J-P and Finne J (1983). Cell adhesion mediated by a purified fucosyltransferase. Proc Natl Acad Sci USA 80:3991.
4. Balsamo J, Pratt RS, Emmerling MR, Grunewald GB and Lilien J (1986). Identification of the chick neural retina cell surface N-acetylgalactosaminyltransferase using monoclonal antibodies. J Cell Biochem 32:125.
5. Taatjes DJ, Roth J, Weinstein J and Paulson JC (1988). Post-Golgi apparatus localization and regional expression of rat intestinal sialyltransferase detected by immunoelectron microscopy with polypeptide epitope-purified antibody. J Biol Chem 263:6302.
6. Shur BD (1989). Expression and function of cell surface galactosyltransferase. Biochem Biophys Acta, in press.

7. Strous GJ (1986). Golgi and secreted galactosyltransferase. CRC Crit Rev Biochem 21:119.
8. Roth J, Lentze MJ and Berger EC (1985) Immunocytochemical demonstration of ecto-galactosyltransferase in absorptive intestinal cells. J Cell Biol 100:118.
9. Pestalozzi DM, Hess M and Berger EG (1982). Immunohistochemical evidence for cell surface and Golgi localization of galactosyltransferase in human stomach, jejunum, liver and pancreas. J Histochem Cytochem 30:1146.
10. Cummings RD, Cebula TA and Roth S (1979). Characterization of a galactosyltransferase in plasma membrane-enriched fractions from Balb/c3T12 cells. J Biol Chem 254:1233.
11. Ebner KE and Mager SC (1975). Lactose synthetase:α-lactalbumin and β1,4 galactosyltransferase. In Ebner K (ed): "Subunit Enzymes. Biochemistry and Function," New York: Marcel Dekker, p. 137.
12. Turley EA and Roth S (1979). Spontaneous glycosylation of glycosaminoglycan substrates by adherent fibroblasts. Cell 17:109.
13. Suganuma T, Muramatsu H, Murata F and Muramatsu T (1987). Purification and properties of N-acetylglucosaminide β1,4 galactosyltransferase from embryonal carcinoma cells. J Biochem 102:665.
14. Strous GJ, Kerkhof PV, Berger EG (1988). In vitro biosynthesis of two human galactosyltransferase polypeptides. Biochem Biophys Res Comm 151:314.
15. Lanford RE, Kanda P and Kennedy RC (1986). Induction of nuclear transport with a synthetic peptide homologous to the SV40 T antigen transport signal. Cell 46:575.
16. Munro S and Pelham HRB (1987). A C-terminal signal prevents secretion of luminal ER proteins. Cell 48:899.
17. Reitman ML and Kornfeld S (1981). UDP-N-acetylglucosamine glycoprotein N-acetylglucosamine-1-phosphotransferase. J Biol Chem 256:4275.
18. Narimatsu H, Sinha S, Brew K, Okayama H and Qasba PK (1986). Cloning and sequencing of cDNA of bovine N-acetylglucosamine β1,4 galactosyltransferase. Proc Natl Acad Sci USA 83:4720.
19. Shaper NL, Shaper JH, Meuth JL, Fox JL, Chang H, Kirsch IR and Hollis GF (1986). Bovine galactosyltransferase: identification of a clone by direct immunological screening of a cDNA expression library. Proc Natl Acad Sci USA 83:1573.
20. Shaper NL, Hollis GF, Douglas JG, Kirsch IR and Shaper JH (1988). Characterization of the full length cDNA for murine β1,4 galactosyltransferase. J Biol Chem 263:10420.
21. Nakazawa K, Ando T, Kimura T and Narimatsu H (1988). Cloning and sequencing of a full length cDNA of mouse N-acetylglucosamine β1,4 galactosyltransferase. J Biochem 104:165.
22. Lopez LC and Shur BD (1988). Comparison of two independent cDNA clones reported to encode β1,4 galactosyltransferase. Biochem Biophys Res Comm 156:1223.

23. Appert HE, Ruthorford TJ, Tarr GE, Wiest JS, Thomford NR and McCorquodale (1986). Isolation of a cDNA coding for human galactosyltransferase. Biochem Biophys Res Comm 139:163.
24. Masri KA, Appert HE and Fukuda MN (1988). Identification of the full length coding sequence for human galactosyltransferase (β-N-acetylglucosaminide: β1,4-Galactosyltransferase). Biochem Biophys Res Comm 157:657.
25. Marchase RB, Kidd VJ, Rivera AA and Humphreys-Beher MG (1988). Cell surface expression of 4β-galactosyltransferase accompanies rat parotid gland acinar cell transition to growth. J Cell Biochem 36:453.
26. Humphreys-Beher MG and Schneyer CA (1987). Cell surface expression of 4β-galactosyltransferase accompanies rat parotid gland hypertrophy induced by changes in diet. Biochem J 246:387.
27. Eckstein DJ and Shur BD (1989). Laminin induces the expression of surface galactosyltransferase on lamellipodia of migrating cells. J Cell Biol, in press.
28. Lopez LC, Maillet CM, Oleszkowicz K and Shur BD (1989) Cell surface and Golgi pools of β1,4 galactosyltransferase are differentially regulated during embryonal carcinoma cell differentiation. Molec Cell Biol, in press.

ALTERED GLYCOSYLATION OF HUMAN CHORIONIC GONADOTROPIN IN TROPHOBLASTIC DISEASES AND ITS USE FOR THE DIAGNOSIS OF CHORIOCARCINOMA[1]

Akira Kobata, Junko Amano, Tsuguo Mizuochi, and Tamao Endo

Department of Biochemistry, Institute of Medical Science, The University of Tokyo, 4-6-1 Shirokanedai, Minato-ku, Tokyo 108, Japan

ABSTRACT

High level of hCG is detected not only in the urine of pregnant women but in those of patients with various trophoblastic diseases. Comparative study of the sugar chains revealed that the oligosaccharide patterns of hCGs purified from urine of pregnant women, patients with invasive mole and patients with choriocarcinoma are different. By making use of the alteration of $\underline{N}$-linked sugar chains, a new diagnostic method to discriminate malignant hCGs from normal hCGs in urine samples has been developed.

INTRODUCTION

Four glycoprotein hormones were found to occur in man. Only human chorionic gonadotropin (hCG) is produced by trophoblasts of placenta and other three hormones are produced in anterior pituitary. Because all of the four hormones are composed of two subunits: α and β, and the amino acid sequences of their α subunit are the same, it had been considered for a long time that each hormone binds to its target cells through its β subunit. However, recent studies of the sugar moieties of these hormones revealed

---

[1]This work was supported by Grant-in-Aid for Scientific Research on Priority Areas (Cancer-Bioscience) from the Ministry of Education, Science and Culture of Japan.

that the α subunits of the four hormones are not the same. Furthermore, the study of the function of the sugar moieties of hCG revealed that the sugars in its α subunit is important for the expression of its hormonal activity.

Structures of the sugar chains of glycohormones was elucidated for the first time on hCG. Each of the two subunits of this hormone contain two N-linked sugar chains. β Subunit contains also four O-linked sugar chains in cluster close to its C-terminal. Structures of N-linked sugar chains of hCG were elucidated independently by Endo et al. (1), and Kessler et al. (2). An interesting and important evidence is that the four asparagine-sites in hCG are glycosylated in a very strict manner (3): the one asparagine site of α subunit has non-fucosylated mono-antennary oligosaccharides and the other non-fucosylated biantennary oligosaccharides, while both of the two asparagine-sites of β subunit contain biantennary oligo-saccharides and the sugar chains of one of them contain a fucose residue (Table 1). The specific distribution of the three groups of oligosaccharides at the four asparagine-loci is considered to be important basis of the functional role of the sugar chains of hCG.

Large amount of hCG is excreted not only in the urine of pregnant women, but in those of patients with various trophoblastic diseases. Structural study of the sugar chains of urinary hCGs obtained from patients with chorio-carcinoma revealed that prominent alteration is induced in their sugar moieties.

TABLE 1
N-LINKED SUGAR CHAINS IN THE TWO SUBUNITS OF HCG

Structures of N-linked sugars	α	β
±NeuSAcα2→3Galβ1→4GlcNAcβ1→2Manα1↘6 Manβ1→4GlcNAcβ1(Fucα1→6)→4GlcNAc NeuSAcα2→3Galβ1→4GlcNAcβ1→2Manα1↗3	−	+
±NeuSAcα2→3Galβ1→4GlcNAcβ1→2Manα1↘6 Manβ1→4GlcNAcβ1→4GlcNAc NeuSAcα2→3Galβ1→4GlcNAcβ1→2Manα1↗3	+	+
Galβ1→4GlcNAcβ1→2Manα1↘6 Manβ1→4GlcNAcβ1→4GlcNAc NeuSAcα2→3Galβ1→4GlcNAcβ1→2Manα1↗3	+	−

Structural Alteration in N-Linked Sugar Chains

Nishimura et al. purified urinary hCG from a patient with choriocarcinoma and found that it gave the same amino acid composition as normal hCG (4). Analysis of the sugar moieties of this hCG sample however revealed that none of the sugar chains was sialylated. Further structural analysis in detail of each oligosaccharide revealed that the

N-I

```
                                              Fucα1
                                                ↓
         Galβ1→4GlcNAcβ1→2Manα1\                6
    A:   Galβ1→4GlcNAcβ1\   6                                          (20.7%)
                          4 Manα1→3 Manβ1→4GlcNAcβ1→4GlcNAc
         Galβ1→4GlcNAcβ1/2

         Galβ1→4GlcNAcβ1→2Manα1\6
    B:   Galβ1→4GlcNAcβ1\4 Manα1→3 Manβ1→4GlcNAcβ1→4GlcNAc               (2.3%)
         Galβ1→4GlcNAcβ1/2
```

N-II

```
                                              Fucα1
                                                ↓
                           Manα1\6              6
    C:   Galβ1→4GlcNAcβ1\4 Manα1→3 Manβ1→4GlcNAcβ1→4GlcNAc              (12.2%)
         Galβ1→4GlcNAcβ1/2

                           Manα1\6
    D:   Galβ1→4GlcNAcβ1\4 Manα1→3 Manβ1→4GlcNAcβ1→4GlcNAc               (9.6%)
         Galβ1→4GlcNAcβ1/2
                                              Fucα1
                                                ↓
         Galβ1→4GlcNAcβ1→2Manα1\6              6
    E:                          Manβ1→4GlcNAcβ1→4GlcNAc                  (7.2%)
         Galβ1→4GlcNAcβ1→2Manα1/3

         Galβ1→4GlcNAcβ1→2Manα1\6
    F:                          Manβ1→4GlcNAcβ1→4GlcNAc                  (4.0%)
         Galβ1→4GlcNAcβ1→2Manα1/3
```

N-III

```
                                              Fucα1
                                                ↓
                           Manα1\6              6
    G:                          Manβ1→4GlcNAcβ1→4GlcNAc                  (8.8%)
         Galβ1→4GlcNAcβ1→2Manα1/3

                           Manα1\6
    H:                          Manβ1→4GlcNAcβ1→4GlcNAc                 (35.2%)
         Galβ1→4GlcNAcβ1→2Manα1/3
```

FIGURE 1. Structures and percent molar ratio of N-linked sugar chains of a choriocarcinoma hCG sample (5).

hormone sample contain the eight neutral oligosaccharides shown in Fig. 1 (5). Note that normal hCG contains oligosaccharides E, F and H but not others. This apparently complicated alteration can be explained by the modification of two glycosyltransferases. Because oligosaccharide G was not detected in normal hCG, expression of the α-fucosyltransferase to form the Fucα1→6GlcNAc group is enhanced. That the total amount of fucosylated oligosaccharides exceeded 50% of the total N-linked sugars, which is twice as that of normal hCG, supported this assumption.

It is evident that oligosaccharides A, B, C and D can be formed from oligosaccharides E, F, G and H, respectively, by the addition of the Galβ1→4GlcNAcβ1→4 group. This structural correlation indicated that β-N-acetylglucosaminyltransferase IV (GnT IV), which forms the GlcNAcβ1→4Man group, is newly expressed in choriocarcinoma. It must be stressed here that oligosaccharides C and D are never found in the normal glycoproteins in human body. Therefore GnT IV in normal tissues does not act on monoantennary oligosaccharide (G and H in Fig. 1), although the enzyme in solubilized form can act on monoantennary oligosaccharide (6). These evidences indicate that GnT IV in the Golgi membrane is controlled by an unknown mechanism, so that it cannot act on the monoantennary complex-type sugar chains. Detection of oligosaccharides C and D in the choriocarcinoma hCG therefore indicated that the ectopically expressed enzyme in this tumor is released from the control mechanism. That the glycosyltransferases in the tumor cells show wider substrate specificities has been noticed in several other tumors. Perhaps the change induced in the Golgi membrane may affect widely to the glycosyltransferases to modify their substrate specificities.

Study of the N-linked sugar chains of four additional choriocarcinoma hCG samples revealed that the lack of sialylation is not always detected in the choriocarcinoma hormone (7). However the eight oligosaccharides in Fig. 1 are always detected in these hormones. Therefore enhanced expression of the fucosyltransferase and the ectopic expression and modification of GnT IV occurs in all choriocarcinoma.

The alteration of N-linked sugar chain was not detected in hCG samples purified from urine of patients with hydatidiform mole. An interesting evidence is that oligosaccharides A, B and G were detected together with E, F and H in the urinary hCG samples purified from patients with invasive mole (8). The oligosaccharides in this case were all sialylated. Invasive mole has been classified as a variant of hydatidiform mole, which shows more malignant characters such as metastasis. Occurrence of oligosaccharides A and B in the hCG produced by this lesion however indicates that it should be considered as a precancerous state.

Structural Alteration in O-Linked Sugar Chains

The data obtained by comparative study of mucin type

TABLE 2
STRUCTURES OF MUCIN-TYPE SUGAR CHAINS OF A NORMAL AND A CHORIOCARCINOMA HCG SAMPLES

Structures	Percent molar ratio	
	normal	choriocarcinoma
Galβ1→4GlcNAcβ1→6 / Galβ1→3 \ GalNAc	2.6	20.6
Galβ1→3GalNAc	27.8	8.7
NeuAcα2→3 { Galβ1→4GlcNAcβ1→6 / Galβ1→3 \ GalNAc }	5.4	23.5
NeuAcα2→6 / Galβ1→3 \ GalNAc	20.4	13.4
NeuAcα2→3Galβ1→3GalNAc	27.1	5.9
NeuAcα2→3Galβ1→4GlcNAcβ1→6 / NeuAcα2→3Galβ1→3 \ GalNAc	2.1	16.0
NeuAcα2→6 / NeuAcα2→3Galβ1→3 \ GalNAc	13.6	4.2

sugar chains in a normal hCG and a choriocarcinoma hCG are summarized in Table 2 (9). As reported by Cole et al. (10), two series of oligosaccharides are detected in normal hCG. These two series contain the Galβ1→3GalNAc and the Galβ1→4GlcNAcβ1→6(Galβ1→3)GalNAc as common cores, respectively. Like in the case of normal hCG, 4 moles of mucin-type sugar chains were detected in choriocarcinoma hCG. An interesting evidence is that the molar ratio of oligosaccharides with the tetrasaccharide core increased prominently in choriocarcinoma hCG. So we investigated the ratio of the two series of mucin-type sugar chains in hCG samples purified from urine of pregnant women and of patients with trophoblastic diseases. The percent molar ratio of oligosaccharides with the tetrasaccharide core in hydatidiform mole hCGs were 8 ~ 11%, which are almost the same as those in normal hCGs (10 ~ 14%). In contrast, those in the hCGs from three choriocarcinoma patients were more than 60%.

Williams et al. (11) reported that the tetrasaccharide core is synthesized by the pathway shown in Fig. 2. Therefore the structural alteration in the O-linked sugar chains of choriocarcinoma hCG should be the result of

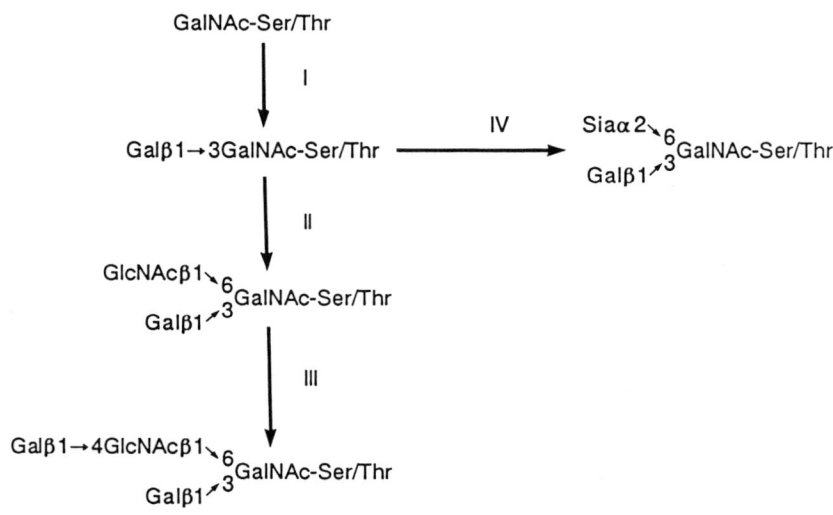

FIGURE 2. Biosynthetic pathway of mucin-type sugar chains (9).

either an enhanced production of $\underline{N}$-acetylglucosaminyltransferase which catalyzes step II in Fig. 2 or suppression of a sialyltransferase which catalyzes step IV and competes with the $\underline{N}$-acetylglucosaminyltransferase for a common acceptor Galβ1→3GalNAcβ1→Ser(Thr) in choriocarcinoma cells. The latter is less unlikely because the total amount of sialylated $\underline{O}$-linked oligosaccharides in normal and choriocarcinoma are almost the same.

Interestingly, the percent molar ratio of oligosaccharides with the tetrasaccharide core in two invasive mole hCGs were 29 and 37%. Therefore the enhancement of the $\underline{N}$-acetylglucosaminyltransferase emerges even at the stage of invasive mole, which is considered to be a precancerous state from the study of $\underline{N}$-linked oligosaccharides of hCG produced by them.

Use of the Altered Glycosylation of HCG for the Diagnosis of Choriocarcinoma

The results so far described indicated that the alteration induced in the $\underline{O}$-linked sugar chains of choriocarcinoma is quantitative while that in the $\underline{N}$-linked sugar chains is qualitative. Because oligosaccharides A and B in Fig. 1 were detected in choriocarcinoma and invasive mole hCGs in common, a method that specifically detects the hCGs containing these oligosaccharides could be used to discriminate these patients from pregnant women and patients with hydatidiform mole.

By investigating the behavior of various complex-type oligosaccharides, liberated from glycoproteins by hydrazinolysis, in a <u>Datura stramonium</u> agglutinin (DSA)-Sepharose column, we found that the oligosaccharides could be separated into three groups: a pass through, a retarded and a bound and eluted with the buffer containing $\underline{N}$-acetylglucosamine oligomers (12). All oligosaccharides recovered in the retarded fraction contain the Galβ1→4GlcNAcβ1→4 (Galβ1→4GlcNAcβ1→2)Man group in non-substituted form. Those which are bound to the column contain either the Galβ1→4GlcNAcβ1→6(Galβ1→4GlcNAcβ1→2)Man group or the Galβ1→4GlcNAcβ1→3Galβ1→4GlcNAc group in non-substituted form. Oligosaccharides which contain none of the groups or contain the groups in either sialylated or fucosylated form pass through the column.

Since oligosaccharides A and B contain the Galβ1→4 GlcNAcβ1→4(Galβ1→4GlcNAcβ1→2)Man group, we expected that

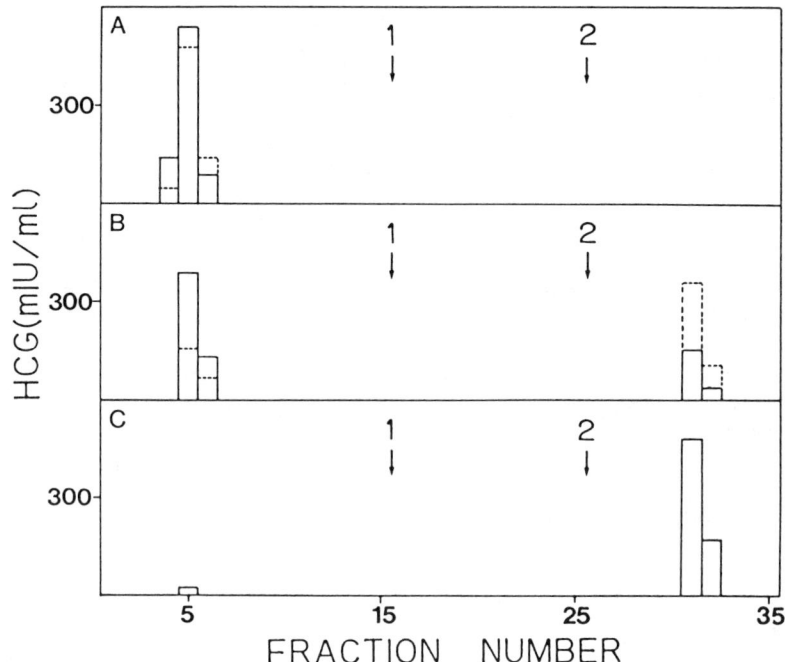

FIGURE 3. Affinity chromatography of urinary hCGs on a DSA-Sepharose column (13). Urine samples, before (solid line) and after (dotted line) sialidase digestion, were applied to a DSA-Sepharose column and the amount of hCG in each fraction was determined by enzyme immunoassay. Arrows 1 and 2 indicate the points where elution buffer was changed to a buffer containing 1% N-acetylglucosamine oligomers and to 0.1 M acetic acid, respectively. A, Urine from a normal pregnant woman; B and C, urine from patients with choriocarcinoma.

the DSA-Sepharose column might be useful for detecting choriocarcinoma and invasive mole hCGs and investigated the behavior of urinary hCGs in this column (13). The Galβ1→4GlcNAcβ1→4(Galβ1→4GlcNAcβ1→2)Man group in the sugar chains of choriocarcinoma hCG may or may not be sialylated. Since the sialylated pentasaccharide residues do not interact with a DSA-Sepharose column, the behavior of hCGs in the urine samples were investigated before and after sialidase digestion. In Fig. 3, elution profiles of two representative choriocarcinoma hCGs and a normal hCG are shown. Almost all hCGs in the urine of a pregnant woman passed through the column (Fig. 3A, solid line). This elution profile did not change after sialidase digestion (Fig. 3A, dotted line). In contrast, only a part of hCG in the urine of a choriocarcinoma patient passed through the column. The remainder was not eluted even with buffer containing N-acetylglucosamine oligomers, but was

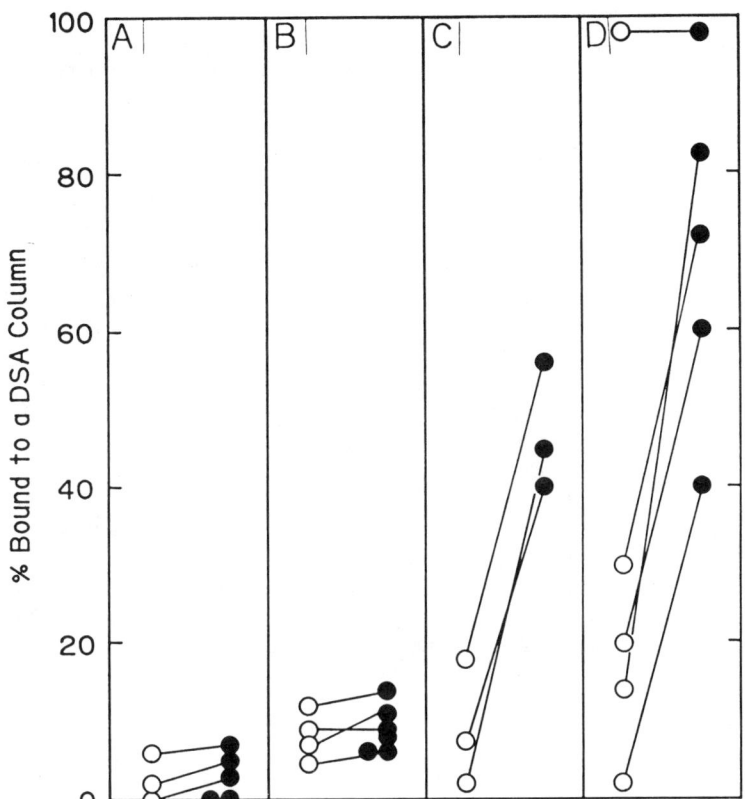

FIGURE 4. Percent molar ratio of urinary hCGs bound to a DSA-Sepharose column before (○) and after (●) sialidase digestion (13). A, Urine samples from normal pregnant women; B, C and D, those from patients with hydatidiform mole, invasive mole and choriocarcinoma, respectively.

completely recovered by elution with 0.1 N acetic acid (Fig. 3B, solid line). This unexpectedly strong binding might be produced because the hCG molecule contains at least two oligosaccharides with the pentasaccharide group interacting with DSA. Therefore the elution step with 1% chitooligosaccharide solution was omitted to obtain the data in Fig. 4. As expected, the portion of hCG bound to a DSA-Sepharose column increased after sialidase digestion (Fig. 3B, dotted line). Most other choriocarcinoma hCGs behaved as shown in Fig. 3B. However, one choriocarcinoma hCG bound completely to the DSA column without sialidase treatment (Fig. 3C, solid line).

After the affinity chromatography of urine samples from five pregnant women, five each of patients with hydatidiform mole and choriocarcinoma and three patients with invasive mole, the percentage of the bound hCG fraction was calculated for each urine sample and the data are summarized in Fig. 4. The values of pregnant women and of patients with hydatidiform mole were less than 15% and the value did not increase after sialidase digestion (Fig. 4A and B). The values of three patients with invasive mole were also low, but became much higher after sialidase digestion (Fig. 4C). These results are in good accord with the evidence that the $\underline{N}$-linked sugar chains of invasive mole hCGs are wholly sialylated although they contain the Galβ1$\rightarrow$4GlcNAcβ1$\rightarrow$4(Galβ1$\rightarrow$4GlcNAcβ1$\rightarrow$2)Man group. The results in the case of choriocarcinoma hCGs showed more variation than in the case of invasive mole hCGs (Fig. 4D). Some of them behaved very similar to invasive mole hCGs. A significant portion of a choriocarcinoma hCG could be bound to the column before sialidase digestion, although the value became much higher after desialylation. One of the choriocarcinoma hCGs bound completely to the column before sialidase digestion as already documented in Fig. 3C. These data accord with the fact that all choriocarcinoma hCG contain the eight oligosaccharide shown in Fig. 1, which are sialylated to various extents.

Concluding Remarks

The data shown in Fig. 4 indicated that the affinity chromatography with a DSA column can be effectively used to discriminate malignant hCGs from normal hCGs in urine samples.

As already discussed, expression of oligosaccharides C

and D is considered to be a real tumor specific phenomenon because these oligosaccharides have never been detected in glycoproteins produced in normal human tissues. We have recently detected these oligosaccharides in γ-glutamyl-transpeptidase produced in hepatocellular carcinoma (14). Therefore the occurrence of these oligosaccharides may not be limited in choriocarcinoma but in other tumors. If it is, the monoclonal antibody directed to these sugars is expected to be useful for the diagnosis of various tumors. With use of a neoglycoprotein synthesized by coupling chemically synthesized abnormal biantennary sugar chain to methylated BSA, we succeeded in producing a monoclonal antibody which specifically binds to abnormal biantennary complex-type sugar chains but not to normal biantennary and triantennary complex type sugar chains. Study with this antibody is expected to develop an useful diagnostic method for tumors.

REFERENCES

1. Endo Y, Yamashita K, Tachibana Y, Tojo S, Kobata A (1979). Structures of the asparagine-linked sugar chains of human chorionic gonadotropin. J Biochem (Tokyo) 85:669.
2. Kessler MJ, Reddy MS, Shah RH, Bahl OP (1979). Structures of N-glycosidic carbohydrate units of human chorionic gonadotropin. J Biol Chem 254:7901.
3. Mizuochi T, Kobata A (1980). Different asparagine-linked sugar chains on the two polypeptide chains of human chorionic gonadotropin. Biochem Biophys Res Commun 97:772.
4. Nishimura R, Endo Y, Tanabe K, Ashitaka Y, Tojo S (1981). The biochemical properties of urinary human chorionic gonadotropin from the patients with trophoblastic diseases. J Endocrinol 4:349.
5. Mizuochi T, Nishimura R, Derappe C, Taniguchi T, Hamamoto T, Mochizuki M, Kobata A (1983). Structures of the asparagine-linked sugar chains of human chorionic gonadotropin produced in choriocarcinoma: appearance of triantennary sugar chains and unique biantennary sugar chains. J Biol Chem 258:14126.
6. Schachter H, Narasimhan S, Gleeson P, Vella G (1983). Enzymatic control of oligosaccharide branching during synthesis of membrane glycoproteins. In Makita A, Tsuiki S, Fujii S, Warren L (eds): "Membrane Altera

tions in Cancer," New York: Plenum Press, p 177.
7. Mizuochi T, Nishimura R, Taniguchi T, Utsunomiya T, Mochizuki M, Derappe C, Kobata A (1985). Comparison of carbohydrate structure between human chorionic gonadotropin present in urine of patients with trophoblastic diseases and healthy individuals. Jpn J Cancer Res (Gann) 76:752.
8. Endo T, Nishimura R, Mochizuki M, Kobata A (1987). Structural differences found in the asparagine-linked sugar chains of human chorionic gonadotropins purified from the urine of patients with invasive mole and with choriocarcinoma. Cancer Res 47:5242.
9. Amano J, Nishimura R, Mochizuki M, Kobata A (1988). Comparative study of the mucin-type sugar chains of human chorionic gonadotropin present in the urine of patients with trophoblastic diseases and healthy pregnant women. J Biol Chem 263:1157.
10. Cole LA, Birken S, Perini F (1985). The structure of the serine-linked sugar chains on human chorionic gonadotropin. Biochem Biophys Res Commun 126:333.
11. Williams D, Longmore G, Matta KL, Schachter H (1980). Mucin synthesis. J Biol Chem 255:11253.
12. Yamashita K, Totani K, Ohkura T, Takasaki S, Goldstein IJ, Kobata A (1987). Carbohydrate binding properties of complex-type oligosaccharides on immobilized Datura stramonium lectin. J Biol Chem 262:1602.
13. Endo T, Iino K, Nozawa S, Iizuka R, Kobata A (1988). Immobilized Datura stramonium agglutinin column chromatography, a novel method to discriminate the urinary hCGs of patients with invasive mole and choriocarcinoma from those of normal pregnant women and patients with hydatidiform mole. Jpn J Cancer Res 79:160.
14. Yamashita K, Totani K, Iwaki Y, Takamizawa I, Tateishi N, Higashi T, Sakamoto Y, Kobata A (1989). Comparative study of the sugar chains of γ-glutamyltranspeptidases purified from human hepatocellular carcinoma and from human liver. J Biochem (Tokyo) in press.

N-GLYCAN PROCESSING AND THE ENTEROCYTIC DIFFERENTIATION OF HT-29 CELLS ARE RELATED EVENTS [1]

Germain Trugnan,* Eric Ogier-Denis,+
C. Bauvy,+ M. Aubery,+ I. Chantret,* C. Sapin,*
P. Codogno,+ and A. Zweibaum *

* INSERM U178, 16 avenue Paul Vaillant-Couturier, 94807 Villejuif-Cedex, France and + INSERM U180 45, rue des Saint-Pères 75006 Paris, France

ABSTRACT A study on the biosynthesis and processing of the cellular glycoproteins has recently shown that undifferentiated HT-29 cells are unable to correctly process their glycoproteins, leading to an accumulation of Man_{9-8}-$GlcNAc_2$ high-mannose glycopeptides. Relations between the enterocytic differentiation and N-glycan processing are further studied here. We have first analyzed the N-glycan processing as a function of the timing of cell differentiation. Data presented in this paper indicate that alterations of N-glycan processing are present very early in cells that will not differentiate. This result led us to conclude that N-glycan processing may be used as an early biochemical probe for the enterocytic differentiation of HT-29 cells. We also analyzed the mechanism by which undifferentiated HT-29 cells accumulate Man_{9-8}-$GlcNAc_2$ species. We show here that the differentiation-dependent changes in the activity of Mannosidase I, an enzyme that converts Man_8-$GlcNAc_2$ into Man_7-$GlcNAc_2$-glycopeptide, are not sufficient to explain the observed alterations of

[1]This work was supported by Association pour la Recherche sur le Cancer (A.R.C.), Villejuif (grant N° 6127), Groupement des Entreprises Françaises de Lutte contre le Cancer (G.E.F.L.U.C.) and Fondation pour la Recherche Médicale (F.R.M.).

N-glycosylation in undifferentiated cells. The use of deoxymannojyrimycin, a specific inhibitor of Mannosidase I, during pulse-chase experiments in both differentiated and undifferentiated HT-29 cells indicates that the stability of the accumulated Man_{9-8}-$GlcNAc_2$ species varies as a function of cell differentiation : these glycopeptides remain stable in differentiated cells whereas they rapidly disappear from undifferentiated cells. These results indicate that the stability of glycoproteins could play a key role in the control of the N-glycans processing as a function of cell differentiation.

INTRODUCTION

Recent data demonstrate that the human colon cancer cell line HT-29 is a very powerful system for the study of the dynamics of cell differentiation as the same cells may exhibit a typical enterocytic differentiation when grown in a glucose-deprived medium or will remain totally undifferentiated when glucose is present during all the time in culture (1-4). It is only when cells grown in a glucose-deprived medium reach confluency that both morphological and enzymatic characteristics of differentiated cells appear, i.e. a polarization of the cell monolayer with an apical brush border, and the presence in the brush border membrane of hydrolase activities, namely sucrase-isomaltase $(SI)^2$, dipeptidylpeptidase IV (DPP-IV), aminopeptidase N (APN), alkaline phosphatase (Alk P) (1-4). Studies on SI, the most specific and sensitive enzymatic marker of the enterocytic differentiation of HT-29 cells have shown that the posttranslational processing of this protein,

[2]Abbreviations Endo H, endo beta-N-acetylglucosaminidase H; DMEM, Dulbecco's modified Eagle's minimum essential medium ; BSA, bovine serum albumin ; Man, mannose ; GlcNAc, N-acetylglucosamine ; HPLC, high performance liquid chromatography ; SI, sucrase-isomaltase ; DPP-IV, dipeptidylpeptidase IV; APN, aminopeptidase N; dMM, deoxymannojyrimycin.
Note : all sugars except fucose are of the D-configuration.

and especially its glycosylation, is impaired in undifferentiated HT-29 cells (5). More recently, we have demonstrated that alterations of protein glycosylation are not restricted to some specific proteins but that the overall N-glycan processing is deficient in confluent undifferentiated HT-29 cells leading to an accumulation of Man_{9-8}-$GlcNac_2$ glycopeptides, whereas differentiated cells exhibit a classical pattern of N-glycosylation (6). Therefore N-glycan processing may be considered as a marker of differentiation, at least in this model.

These results raised two questions that were studied in this paper. (i) Are these alterations of the N-glycan processing present from the beginning of the culture or do they appear progressively during the cell growth and (ii) what is the mechanism responsible for the blockade of the trimming of Man_{9-8}-$GlcNAc_2$ species in undifferentiated HT-29 cells. The first question was studied by comparing four distinct cell populations : (a) growing HT-29 Glc⁻ cells having an undifferentiated phenotype; (b) confluent HT-29 Glc⁻ cells which are differentiated; (c) and (d) growing and confluent HT-29 Glc⁺ cells that remain undifferentiated. Pulse-chase experiments using 2-(^{3}H) mannose were performed and showed, in HT-29 Glc⁺ cells only, an altered N-glycan processing that does not depend on the phase of growth. In contrast, HT-29 Glc⁻ cells, even during the exponential phase of growth display an almost normal pattern of N-glycosylation. From these results we conclude that the N-glycan processing may be used as an early biochemical probe for the enterocytic differentiation of HT-29 cells. To answer the second question, the activity of mannosidase I was measured in both differentiated and undifferentiated cells using an in vitro protocol (7,8). It was shown that although there are some differences in the kinetic parameters of the enzyme as a function of the state of differentiation, mannosidase I is active in both conditions and therefore cannot be directly responsible for the accumulation of Man_{9-8}-$GlcNac_2$ species in undifferentiated cells. The use of dMM, a known inhibitor of mannosidase I (9,10), demonstrates that the the fate of high-mannose species that accumulate differs considerably as a function of the state of enterocytic differentiation: these glycopeptides remain stable for more than 24 hours in differentiated cells, whereas they rapidly disappear in undifferentiated cells. These results suggest the presence in undifferentiated HT-29 cells of a degradation mechanism for unprocessed high mannose glycopeptides similar to what has been recently described (11).

MATERIALS AND METHODS

Cells and culture conditions.

HT-29 cells (12) were cultured as previously reported (5). The same cells were grown in a medium containing 25 mM glucose (HT-29 Glc$^+$ cells : differentiation-non-permissive conditions) or without glucose containing 2.5 mM Inosine (HT-29 Glc$^-$ cells : differentiation-permissive conditions). The cells were used after confluency (18-20 days) when HT-29 Glc$^-$ cells exhibit a typical enterocytic differentiation and during the early phase of growth (5 days) when the differentiation criteria are not yet expressed in HT-29 Glc$^-$ cells (2,4).

Preparation of cellular extracts and enzyme assays.

Cell homogenates and crude membrane fractions were prepared according to methods described elsewhere (13). Sucrase (EC 3.3.1.48), aminopeptidase N (EC 3.4.11.2) and dipeptidylpeptidase IV (EC 3.4.14.5) activities were measured in the cell homogenate according to previously described methods (14-16). The activities are expressed as milliunits (mU) per mg of proteins. One unit is defined as the enzymatic activity which hydrolyzes 1 µmole of substrate per min at 37°C. Mannosidase I was assayed using the oligosaccharide Man$_8$-GlcNAc$_1$ as substrate. 6.5 µM to 12.5 µM of substrate in the presence of the same tritiated mannose labeled oligosaccharide (10,000 cpm) were incubated at 37°C from 0 to 120 min with crude microsomal fractions obtained from each cell population in the presence or absence of inhibitors (7,8). Tritiated mannose and the remaining oligosaccharides were separated using Concanavalin A (17).

Cell labeling and glycopeptide chromatography.

Metabolic labeling of the cells was performed using D-(2-^{3}H)-mannose (20 Ci/mmol, The Amersham Radiochemical

Centre, U.K.) for either short (10 min, 400 μCi/ml pulse experiments) or long (24h, 40 μCi/ml) periods as previously described (6). However, it was shown that cancer cells very actively convert mannose into fucose (18 and table 1 below). Therefore 2mM fucose was added to the chase medium and this was shown to prevent this metabolic conversion (figure 1). Extraction (19) and purification (20) of the labeled glycopeptides were as previously described. Endo H digestion was performed according to Tarentino and Maley (21).

TABLE 1
CONVERSION OF D-[2-^{3}H]-MANNOSE INTO [^{3}H]-FUCOSE

| | HT-29 Glc$^-$ | | HT-29 Glc$^+$ | |
	Growing	Confluent	Growing	Confluent
Fraction I[a]	57[b]	67	66	66
Fraction II	69	49	50	48
Fraction III	60	41	63	47
Fraction IV	0	0	0	0

a - Growing and confluent HT-29 Glc$^-$ and HT-29 Glc$^+$ cells were labeled for 24h and the different glycopeptides were resolved by a Bio-Gel P6 chromatography.

b - Results are expressed as the percentage of radioactivity found in fucose after a 24h labeling period using D-[2-^{3}H]-mannose as precursor.

Oligosaccharides released by endo H treatment were fractionated by HPLC in a Varian model 5000 liquid chromatograph equipped with a column of 5 μ aminospherisorb (Société Française de Chromatographie, France) as previously reported. Oligosaccharides were fractionated in the presence of ^{14}C labeled oligosaccharide standards (Figure 1).

Conversion of (^{3}H)-mannose.

Tritiated glycopeptides were hydrolyzed either with 0.05 M H_2SO_4 for 8h at 100°C or with glacial acetic acid and 5 M H_2SO_4 in a ratio 95:5 (v/v) for 4h at 80°C. The recovery of fucose was similar under both hydrolytic conditions. After desalting, both hydrolysates were spotted and run for 40h on Schleicher and Schüll 2043b paper in 1-butanol/ pyridine/0.1 M HCl (5:3:2, by vol.).

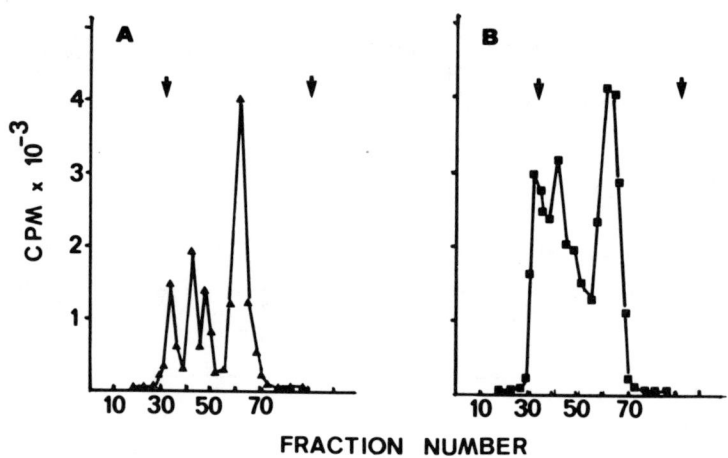

FIGURE 1. Bio-Gel P6 chromatography from confluent HT-29 Glc⁻ cells labeled for 24h either in the presence (A) or in the absence (B) of 2 mM fucose. Arrow indicate the elution of BSA and mannose respectively.

RESULTS

1. The growth and differentiation characteristics of HT-29 cells allow studies on the early events of enterocytic differentiation.

HT-29 cells were grown in DMEM containing 2.5 mM inosine without (differentiation-permissive conditions) or with (differentiation- non-permissive conditions) 25 mM glucose. Growth curves were established on these two cell populations and are displayed in Figure 2. In the same experiment, enzymatic activities were measured as described in the Method Section. In HT-29 Glc⁻ cells the activities of DPP-IV, APN (not shown), and sucrase (Figure 2) which were absent or low in the exponential phase of growth progressively increased after the cells had reached confluency. No such increase could be detected in HT-29 Glc+ cells (23) (Figure 2).

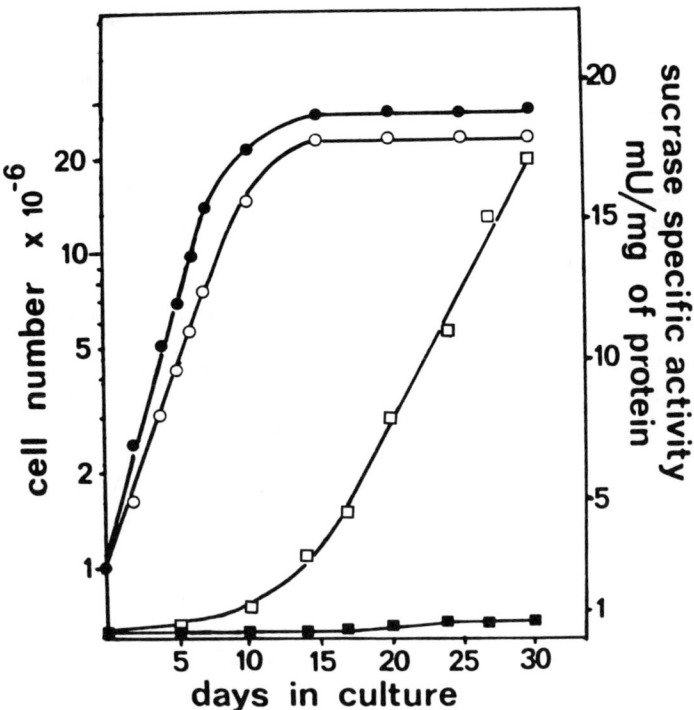

FIGURE 2. Growth curve (circle) and sucrase specific activity (square) in HT-29 Glc− (open symbols) and HT-29 Glc+ (closed symbols) cells as a function of the time in culture. Other brush border associated enzymes (not shown) display the same pattern as sucrase.

Immunofluorescence and transmission electron microscopy studies have shown that HT-29 cells are not polarized during the 6-7 first days in culture whatever the glucose content of the culture medium (4). However, a functional brush border appears after 10 days in HT-29 Glc− cells, whereas HT-29 Glc+ cells remain unpolarized (data not shown).

Such a modulatable cellular system allows to predict how the cells will be differentiated after confluency eventhough they are undistinguishable, on the basis of the reported criteria, during the exponential phase of growth. We have therefore defined four distinct cell populations : growing and confluent HT-29 Glc$^-$ and growing and confluent HT-29 Glc$^+$ cells that allow to describe the change of N-glycan processing as a function of both the phase of growth and the differentiation state of the cells.

2. Undifferentiated HT-29 Glc$^+$ cells accumulate high mannose species during the exponential and the stationary phases of growth.

After a 24h labeling period using D-(2-^{3}H)-mannose, glycopeptides from HT-29 Glc$^-$ and HT-29 Glc$^+$ cells were extensively digested with pronase and fractionated on a column of Bio-Gel P6 (Figure 3). In the four cell populations 4 peaks were resolved. In each case only peak IV was sensitive to endo-H treatment. Furthermore this peak exhibited a high affinity for Con A Sepharose, whereas peaks I, II and III were not retained on this immobilized lectin (data not shown). From the results presented in figure 3 it appeared that whatever the situation considered (i.e. growth or confluency) HT-29 Glc$^+$ cells exhibited a higher percentage of radioactivity in high mannose fraction than HT-29 Glc$^-$ cells (Figure 3).

3. The alteration of high mannose processing is an early event in undifferentiated HT-29 Glc$^+$ cells.

In order to explain the accumulation of high mannose species in HT-29 Glc$^+$ cells, the N-glycan processing was followed in the four cell populations using pulse-chase experiments. After a 10 min pulse with D-(2-^{3}H)-mannose, the cells were incubated with 5 mM unlabeled mannose and 2 mM fucose for various periods of time. At each time the radioactivity associated with lipid-linked oligosaccharides, high mannose glycopeptides, and complex glycopeptides was measured (Figure 4 a,b). The nature of the radioactivity associated with labeled glycopeptides was analyzed and was found to be exclusively associated with mannose. In HT-29 Glc$^-$ cells

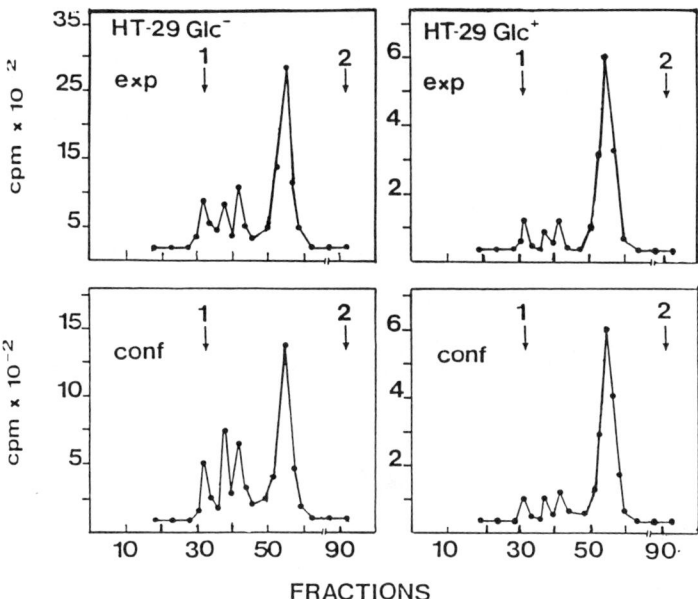

FIGURE 3. Bio-gel P6 chromatography of mannose-labeled glycopeptides from growing and confluent HT-29 Glc$^-$ and HT-29 Glc$^+$ cells. Cells were labeled for 24h in the presence of 2 mM fucose that prevent the conversion of [2-^{3}H]-mannose into [^{3}H]-fucose. Note the low level of complex glycopeptides in growing and confluent HT-29 Glc$^+$ cells.

the fate of ^{3}H-mannose was similar to what expected for a classical pathway of N-glycosylation (24) (Figure 4a).
The biosynthesis of labeled lipid-linked oligosaccharides and their transfer to high mannose species was qualitatively similar in all the situations tested. Incontrast a low level of radioactivity was recovered in complex N-glycans from HT-29 Glc$^+$ as compared to HT-29 Glc$^-$ cells. Therefore the ratio : complex/high mannose N-glycans was lower in HT-29 Glc$^+$ than in HT-29 Glc$^-$ cells (Figure 4). These results indicate that whatever the phase of growth, the complex N-glycan biosynthesis is similarly impaired in HT-29 Glc$^+$ cells. In contrast the

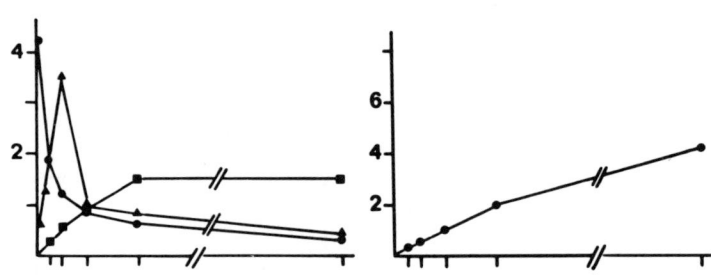
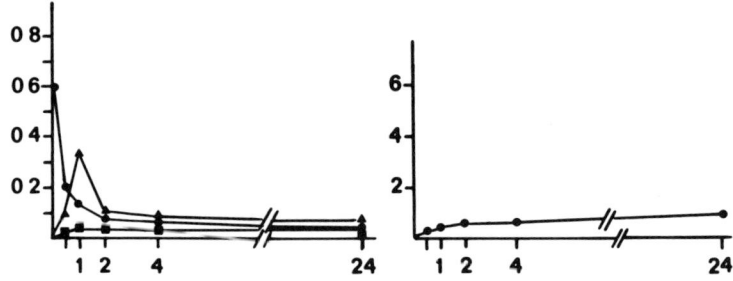

FIGURE 4a. Pulse-chase experiment on growing HT-29 Glc⁻ and HT-29 Glc⁺ cells. Cells were labeled at day 5 using [2-^{3}H]-mannose and lipid-linked oligosaccharides (●—●), high-mannose (▲—▲) and complex glycopeptides (■—■) were analyzed as described in the material and methods section. Vertical axis of the left pannel correspond to the recovered cpm (cpmx10^{-3}) and vertical axis of the right pannel correspond to the ratio of the radioactivity in complex and in high mannose glycopeptides.

biosynthesis of complex N-glycans is not altered in either growing or post-confluent HT-29 Glc⁻ cells (Figure 4b).

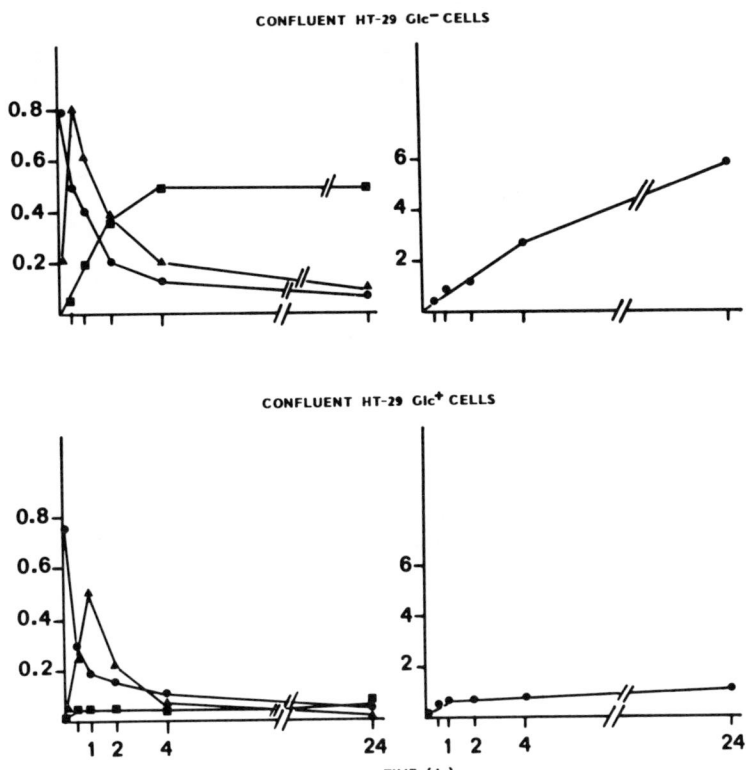

FIGURE 4b. Pulse-chase experiment on confluent HT-29 Glc- and HT-29 Glc+ cells. Legend as in Fig. 4a.

High mannose species were studied in the endo H sensitive fractions from the 24h labeling experiments. Results of HPLC analysis of the different high mannose types are displayed in Table 2. These results clearly demonstrate that both growing and confluent HT-29 Glc+ cells accumulate Man_{9-8}-$GlcNAc_2$ species, whereas HT-29 Glc- exhibit a more uniform oligosaccharide distribution. It should be noted that growing HT-29 Glc- cells have an higher proportion of Man_{9-8}-$GlcNac_2$ compared to their confluent counterpart. However in this case, this is not associated to a blockade in the processing of these glycopeptides (Figure 4). Therefore the mechanism underlying the impairment of N-glycan trimming in growing HT-29 Glc+ cells seems to be due to an accumulation of Man_{9-8}-$GlcNAc_2$ glycopeptides, as previously shown in confluent HT-29 Glc+ cells.

4. Mannosidase I is active in both differentiated and undifferentiated cells.

In both differentiated and undifferentiated HT-29 cells the activity of mannosidase I toward Man_8-$GlcNAc_1$ oligosaccharide was detected. The kinetic parameters of the enzyme were determined from the Linewaever and Burk representation (Figure 5). The Km was in a 12-17 μM range for differentiated cells and 35-52 μM for undifferentiated cells. These values were in good agreement with data from rat liver mannosidase I (7). Similarly the calculated Vmax values were in the same range : 9 μmol/min/mg of protein and 18 μmol/min/mg of protein for differentiated and undifferentiated cells respectively. These results showed that the kinetic parameters were very close in both cell populations and therefore cannot explain the difference observed in the trimming of high mannose glycoproteins. Furthermore these data allowed us to investigate the N-glycan processing in the presence of dMM to follow the fate of N-glycan chains when blocked at the same step of trimming (Figure 5).

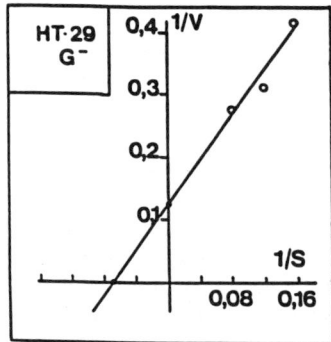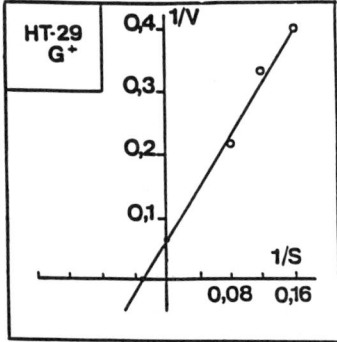

FIGURE 5. Linewaever and Burk representation of the alpha-mannosidase I activity in confluent HT-29 Glc− and HT-29 Glc+ cells. Kinetic parameters derived from this figure are given in the text.

5. High mannose glycopeptides are rapidly degraded in undifferentiated cells.

In a preliminary set of experiments we have shown that in the presence of dMM, all the mannose labeled glycopeptides were sensitive to endo H treatment and that as expected both differentiated and undifferentiated cells accumulate Man_{9-8}-$GlcNac_1$ oligosaccharides (data not shown). Pulse chase experiments in the presence of dMM were performed as detailed in Figure 3. In the presence of the inhibitor, a quantitative transfer of radioactivity is observed from the lipid donor to the high mannose polypeptides (Figure 6). However, in differentiated cells the level of radioactivity associated with thes high mannose chains remained high and constant all along the chase period, whereas in undifferentiated cells there is a rapid decrease with almost no more radioactivity detectable after 2 hours in the high mannose glycopeptides. These results suggest that the impairment of N-glycan processing in undifferentiated HT-29 cells could be correlated to a rapid degradation, avoiding a further processing of oligosaccharide chains (Figure 6).

DISCUSSION

The results presented in this paper demonstrate that the N-glycan processing is independent of the phase of growth of HT-29 cells but strongly depends on the ability of cells to differentiate. Therefore it can be concluded that although the two cell populations are indistinguishable during the exponential phase of growth according to morphological and enzymatic criteria it is however possible to distinguish them by using a biochemical tool, namely the N-glycan processing. Therefore, at least in HT-29 cells, the N-glycan processing appears to be a sensitive and early biochemical probe of the ability of these cells to differentiate.

In these cells the step that is defective is the conversion of Man_{9-8}-$GlcNAc_2$ into Man_{7-5}-$GlcNAc_2$ species. This could result either from the defect of Mannosidase I on from the unability of the substrate to reach the enzyme. Direct measurements of the enzyme activity show that there are no significant variations as a function of the differentiation state of the cells. In contrast experiments using dMM indicate that there is a rapid

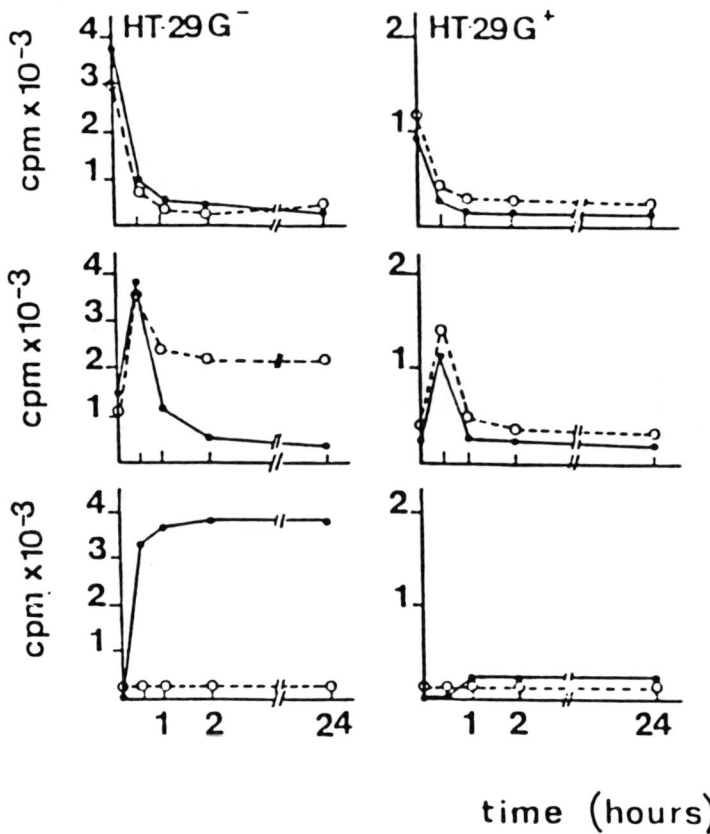

FIGURE 6. Pulse-chase experiment on HT-29 Glc$^-$ and HT-29 Glc$^+$ cells in the absence (closed symbols) or in the presence (open symbols) of deoxymannojirimycine. The experiment was carried out as in Fig. 4a. Upper pannel : lipid-linked oligosaccharides ; middle pannel : high mannose glycopeptides ; lower pannel : complex glycopeptides. Note the absence of complex glycopeptides in both cell populations when treated with dMM and the differential stability of the high-mannose glycopeptides.

degradation of high-mannose type chains in undifferentiated cells. Whereas in differentiated cells these molecules remain stable for at least 24 h. These observations may be part of the explanation as one can easily hypothesize that in undifferentiated cells high-mannose chains cannot be further processed only because they do not reach the enzyme. The mechanism by which high-mannose glycoproteins rapidly disappear in undifferentiated cells is still unknown. However some recent papers have suggested the possibility of a very precocious degradation for unprocessed or malfolded proteins and a recent work has demonstrated the presence of a pre-Golgi degradative pathway in T-cells (11). Whether this process is involved here and could be related to the cell differentiation remains to be studied.

REFERENCES

1. Pinto M, Appay MD, Simon-Assmann P, Chevalier G, Dracopoli N, Fogh J, Zweibaum A (1982) Enterocytic differentiation of cultured human colon cancer cells by replacement of glucose by galactose in the medium. Biol Cell 44:193.
2. Wice BM, Trugnan G, Pinto M, Rousset M, Chevalier G, Dussaulx E, Lacroix B, Zweibaum A (1985) The intracellular accumulation of UDP-N-acetylhexosamines is concomitant with the inability of human colon cancer cells to differentiate. J Biol Chem 260:139.
3. Zweibaum A, Laburthe M, Grasset E, Louvard D (1989). In Field M, Frizzel A (eds): "Handbook of Physiology. Intestinal Transport of the Gastrointestinal System". Am Physiol Soc (In press).
4. Zweibaum A, Pinto M, Chevalier G, Dussaulx E, Triadou N, Lacroix B, Haffen K, Brun JL, Rousset M (1985). Enterocytic differentiation of a subpopulation of the human colon tumor cell line HT-29 selected for growth in sugar-free medium and its inhibition by glucose. J Cell Physiol 122:21.
5. Trugnan G, Rousset M, Chantret I, Barbat A, Zweibaum A (1987). The post-translational processing of sucrase-isomaltase in HT-29 cells is a function of their state of enterocytic differentiation. J Cell Biol 104:1199.
6. Ogier-Denis E, Codogno P, Chantret I, Trugnan G (1988). The processing of asparagine-linked oligosaccharides in HT-29 cells is a function of their state of enterocytic differentiation. An accumulation of Man_{9-8}-Glc-NAc_2-Asn species is concomitant with an impaired N-glycans trimming. J Biol Chem 263:6031.

7. Tabas I, Kornfeld S (1979). Purification and characterizaton of a rat liver Golgi alpha-mannosidase capable of processing asparagine-linked oligosaccharides. J Biol Chem 254:11655.
8. Tulsiani DRP, Hubbard SC, Robbins PW, Touster O (1982). Alpha-D-mannosidases of rat liver Golgi membranes. Mannosidase II is the GlcNAcMan$_5$-cleaving enzyme in glycoprotein biosynthesis and mannosidases IA and IB are the enzymes converting man$_9$ precursors to Man$_5$ intermediates. J Biol Chem 257:3660.
9. Fuhrmann U, Bause E, Legler G, Ploegh H (1984). Novel mannosidase inhibitor blocking conversion of high mannose to complex oligosaccharides. Nature 307:755.
10. Bischoff J, Kornfeld R (1984). The effects of 1-deoxymannojirimycine on rat liver alpha-mannosidase. Biochem Biophys Res Comm 125:324.
11. Lippincot-Schwartz J, Bonifacino JS, Yuan LC, Klausner RD (1988). Degradation from the endoplasmic reticulum : disposing of newly synthesized proteins. Cell 54:209.
12. Fogh J, Trempe G (1975). In Fogh J (ed): "Human Tumor Cells in vitro", New York: Plenum Publish Corp, p 115.
13. Schmitz J, Preiser H, Maestracci D, Ghosh BK, Cerda JJ, Crane RK (1973). Purification of the human intestinal brush border membrane. Biochim Biophys Acta 323:98.
14. Messer M, Dahlqvist A (1966). A one-step ultra micromethod for the assay of intestinal dissacharides. Anal Biochem 14:376.
15. Maroux S, Louvard D, Baratti J (1973). The aminopeptidase from hog intestinal brush border. Biochim Biophys Acta 321:282.
16. Nagatsu T, Hino M, Fuyamada H, Hayakawa T, Sakakibara S, Nakagawa Y, Tatemoto T (1976). New chromogenic substrates for X-prolyl-dipeptidyl-aminopeptidase. Anal Biochem 74:466.
17. Herskovics A, Jelinek-Kelly S (1987). A rapid method for the assay of glycosidases involved in glycoprotein biosynthesis. Anal Biochem 166:85.
18. Youakim A, Herscovics A (1985). Cell surface glycopeptides from human intestinal epithelial cell lines from normal colon and colon adenocarcinomas. Cancer Res 45:5505.

19. Cummings RD, Kornfeld S (1982). Fractionation of asparagine-linked oligosaccharides by serial lectin- agarose affinity chromatography. A rapid, sensitive, and specific technique. J Biol Chem 257:11235.
20. Codogno P, Botti J, Font J, Aubery M (1985). Modification of the N-linked oligosaccharides in cell surface glycoproteins during chick embryo development. A using lectin affinity and a high resolution chromatography study. Eur J Biochem 149:453.
21. Tarentino AL, Maley F (1974). Purification and properties of a endo-beta-N-acetylglucosaminidase from Streptomyces griseus. J Biol Chem 249:811.
22. Romero PA, Saunier B, Herscovics A (1985) Comparison between 1-deoxymannojirimycine and methyl-1-deoxy- jirimycine as inhibitors of oligosaccharides processing in intestinal epithelial cells. Biochem J 226:733.
23. Ogier-Denis E, Bauvy C, Aubéry M, Codogno P, Sapin M, Rousset M, Zweibaum A, Trugnan G (1989). The processing of asparagine-linked oligosaccharides is an early biochemical marker of the enterocytic differentiation of HT-29 cells. J Cell Biochem (in press).
24. Kornfeld R, Kornfeld S (1985). Assembly of asparagine linked oligosaccharide. Ann Rev Biochem 256:631.

NUCLEAR PORE GLYCOPROTEINS: STRUCTURE AND FUNCTION

John A. Hanover, Min Kyun Park, Mara D' Onofrio, Christopher Starr, Tracy S. Olson and Barbara Wolff

Laboratory of Biochemistry and Metabolism, NIDDK, National Institutes of Health Bethesda, MD 20892

ABSTRACT The objective of the chapter is to integrate knowledge of nuclear pore structure with recent information regarding the function of the pore in nuclear protein import and RNA efflux. First, studies leading to the identification of the glycoprotein components will be described. Secondly, the role of the pore proteins in nuclear protein import will be addressed. The involvement of the pore complex in RNA transport will then be briefly described. Finally, information obtained from the molecular cloning of one of the nuclear pore glycoproteins will be outlined.

INTRODUCTION

The nuclear membrane or nuclear envelope is very different from any other membrane system in the eukaryotic cell (1,2). It consists of two lipid bilayers enclosing a perinuclear cisternal space. The outer membrane is continuous with the endoplasmic reticulum and has attached ribosomes, while the inner nuclear membrane may form attachment points for chromatin through the nuclear lamins. The nuclear envelope is interrupted at various intervals by pores which form a morphological connection between nucleus and cytoplasm (3,4). There is evidence to suggest that nucleo-cytoplasmic transport occurs through the aqueous channels of the nuclear pore. The pores allow passive diffusion of electrolytes and other small molecules (5,6) as well as some macromolecules. However, dextrans with a MW of more than 22,000 and globular proteins with a MW of more than 65,000 are excluded from the nucleus suggesting a functional radius of the pore of 35Å.

It has become clear that large proteins require specific localization sequences for their proper transport across the nuclear envelope. A number of such sequences that can act independently to confer nuclear localization have been identified. For most of these proteins, however, it cannot yet be determined unequivocally whether their overall shape or specific polypeptide sequences are involved in the transport. Furthermore, the specific mechanism for this translocation process is poorly understood. Growing evidence suggests that the most likely place for signal recognition by a "receptor" is the nuclear pore itself.

As is the case for protein import, simple diffusion cannot account for the export of ribonucleoprotein particles and RNA from the nucleus. RNA leaves the nucleus only in a mature, fully processed form. The transport event is energy dependent; it involves release of the RNA from some component of the nuclear substructure and then a bidirectional export event. The mechanism of these steps, while under active investigation, is poorly understood. While nuclear pore complexes seem to be involved in RNA transport, it is not clear whether the same pores are involved in both RNA transport <u>out of</u> and protein uptake <u>into</u> the nucleus.

IDENTIFICATION OF NUCLEAR PORE GLYCOPROTEINS

While most of the current information about the structure of the nuclear pore complex has come from morphological examination, biochemical characteristics of some of the proteins comprising the pore complex have been recently elucidated. Critical to the identification of these proteins has been the finding that they contain a novel carbohydrate modification consisting of O-linked N-acetylglucosamine (GlcNAc) covalently attached to Ser and Thr residues of the polypeptide (7,8,9). A lectin, wheat germ agglutinin, binds to these proteins by associating with the O-linked N-acetylglucosamine residues (8). WGA-ferritin was found to decorate the pore complex of isolated rat liver nuclei (Fig 1.)(8). Monoclonal antibodies against nuclear pore proteins have been developed in several laboratories including our own

(10,11,12). These antibodies recognize a family of

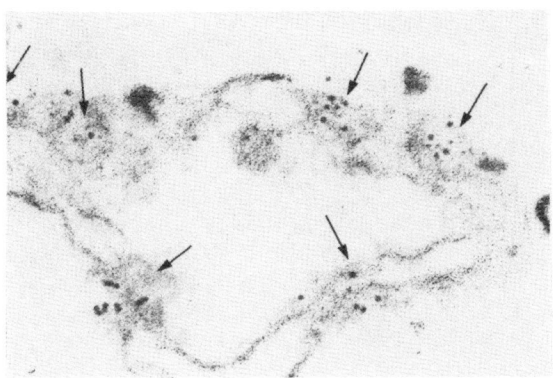

FIGURE 1. WGA-ferritin binding to nuclear pore complexes of rat liver. Arrows indicate the position of pore complexes.

proteins which all seem to bear O-linked N-acetylglucosamine and which have apparent molecular weights ranging from 45,000 to over 200,000. Surprisingly, removal of the N-acetylglucosamine residues has been shown to destroy the antigenicity of these proteins, strongly suggesting that O-linked GlcNAc is part of the immunodeterminant recognized. Immunoelectron microscopy has demonstrated that the antigens recognized by the antibodies are located exclusively within the nuclear pore complex at both the nucleoplasmic and cytoplasmic surfaces. When examined at the level of light microscopy by immunofluorescence, the antibodies give a highly punctate pattern.

INVOLVEMENT OF THE NUCLEAR PORE IN NUCLEAR PROTEIN UPTAKE

To date, the localization sequence that has been best characterized is the SV40 large T antigen sequence.

Large T antigen is a DNA binding protein with a subunit molecular weight of 94,000. It is too large to enter the nucleus passively. Recombinant DNA techniques have allowed a clear dissection of the sequences responsible for Large T antigen nuclear localization. For example, the conversion of a Lys within the sequence Pro-Lys-Lys^{128}-Lys-Arg-Lys-Val into any number of amino acids completely abolished nuclear localization (13,14). The necessary and sufficient amino acids that could act as an autonomous signal to confer nuclear localization to chimaeric proteins like ß-galactosidase or pyruvate kinase were either amino acids 128-135 or 126-132 (15). This sequence dependence was also demonstrated using proteins which were chemically coupled to synthetic peptides containing the SV40 T antigen amino acids 126-132. Changing the Lys residue corresponding to Lys ^{128} to either Thr or Asn prevented or significantly reduced nuclear uptake of the conjugates (16,17,18,19,20). Also, sequentially deleting amino acids from the C-terminus of the peptide confirmed the requirement for the sequence up to at least Val ^{132} (20).

To determine how nuclear localization sequences may be recognized, we have applied an immunological approach. Although the position of the nuclear targeting signal in the protein molecule is not necessarily the amino- or carboxy terminus, there is reason to believe that the sequence needs to be exposed on the surface of the molecule. If there were a "receptor" for nuclear localization sequences, the sequence would have to be available for interaction with a binding site. Studies using the SV40 T antigen squence have revealed that it could be introduced into several different sites within pyruvate kinase and still function correctly (21). However, when it was inserted into a part of the molecule that was buried according to crystallographic studies, the signal was inactive. A differential function depending on the location within the protein molecule would explain why sequences resembling the SV40 T polypeptide occur in a number of cytoplasmic proteins as well (22). Monoclonal antibodies against the localization signal of SV40 bind to in vitro translated SV40 T antigen and to the T antigen expressed in cultured cells (Wolff, Park & Hanover, manuscript in preparation). These data also strongly suggest that the localization sequence is in an exposed portion of the molecule. To establish an assay

to directly examine nuclear transport we have developed a
means of coupling the SV40 Large T antigen nuclear
localization sequence to the highly fluorescent molecule
phycoerythrin (20). The conjugation results in the
formation of a thioether bond which is stable in the
cytoplasm of cells. Direct microinjection of such
conjugates into the cytoplasm of fused cells provides a
convenient assay for transport (Fig 2.). In living cells,
the transport of such conjugates is both time and

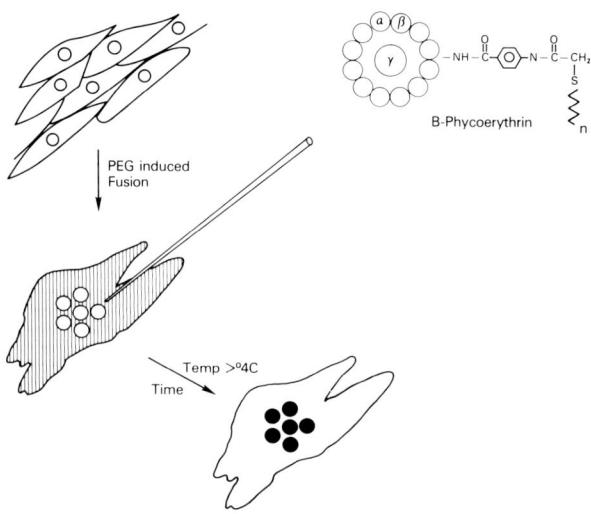

FIGURE 2. Microinjection assay for nuclear localization
sequences.

temperature dependent (20) and is complete within 30
minutes of injection. To determine whether the lectin WGA
would interfere with nuclear transport, 3T3 cells were
microinjected with a combination of WGA (5mg/ml) and the
B-phycoerythrin-T peptide conjugate described above.
Instead of accumulating in the nucleus within the first 30
minutes, the fluorescent protein remained in the cytoplasm
for 3 hours, until the uptake block was reversed by
microinjection of the competing sugar. The degree of
inhibition of uptake was dependent on the lectin

concentration (20,23). In spite of its binding to nuclear pores, WGA did not sterically block the accumulation of all macromolecules in the nucleus; dextrans with a molecular weight of 10,000-18,000 could still passively diffuse into the nucleus (19,20,23).

An in vitro assay for nuclear transport has also been established. Using the phycoerythrin conjugates described above, rat liver nuclei can be demonstrated to accumulate the conjugate in the nuclear interior (Fig 3). This accumulation is not observed when an inactive peptide derived from a cytoplasmic Large T antigen (cT) is

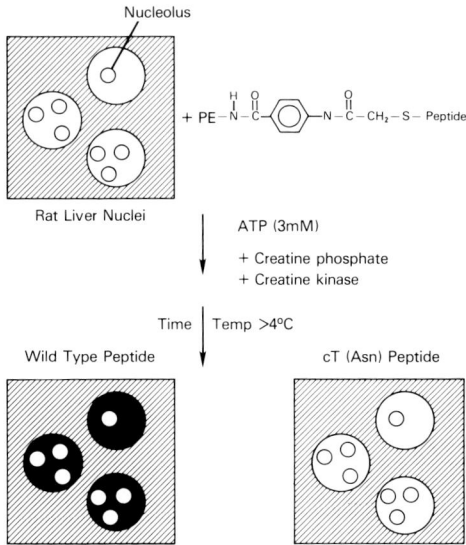

FIGURE 3. In vitro assay for nuclear protein Transport.

coupled to phycoerythrin. The uptake of the active conjugate is dependent upon the inclusion in the incubation of ATP (optimal concentration of 3mM). Tranport is not observed at 4 ° C nor in the presence of WGA. This assay is currently being modified to obtain quantitative information concerning the interaction between such conjugates and the nuclear membrane.

INVOLVEMENT OF THE PORE IN RNA TRANSPORT FROM THE NUCLEUS

All cytoplasmic RNA species must be transcribed from the DNA molecules present in the nucleus. The RNA species then undergo a series of post-transcriptional modifications which include processing, splicing and polyadenylation-all thought to be nuclear events. The RNA is then ultimately transported to the cytoplasm. Although RNA species are single stranded, they may have substantial secondary structure due to double helical regions arising from folding of the chain into hairpins. How much this secondary structure contributes to their recognition, processing and transport is yet to be determined. Recently, the translocation of mRNA and tRNA species from nucleus to cytoplasm have been investigated. The tRNAs are the smallest RNA molecules, containing between 73 and 93 nucleotides (about 25 kdal). Molecules of this size might be small enough to diffuse through the nuclear pore passively. However the recent identification of transport defective mutants strongly argues against passive diffusion. A single G to U substitution at position 57 in the vertebrate $tRNA^i$ <Met> molecule was found to reduce the transport rate of this tRNA by a factor of about 20. The highly conserved region of the tRNA molecule was interpreted to be critical for recognition by the transport mechanism (24). The mechanism by which a tRNA molecule is delivered from the nucleus to the cytoplasm has been studied in the oocyte of *Xenopus laevis* using nuclear microinjection and manual microdissection techniques (Fig 4). In this system, tRNA nuclear transport resembles a saturable, carrier-mediated translocation process rather than diffusion through a simple pore or channel (25). Zasloff and coworkers have proposed that ribosome-like components surrounding the nuclear pore may function as the tRNA translocation "motor". However, no direct experiments have been done to test this model.

In the experiments we have performed, the lectin WGA was introduced into the nucleus with the appropriate tRNA species (tRNA Met). In some cases, competing saccharide was added to prevent lectin binding to the pore complex. In a series of experiments of this kind, we have found that the lectin inhibits the rate of tRNA transport by greater than 50%. Succinylated WGA, which has a much higher affinity for GlcNAc than for sialic acid residues

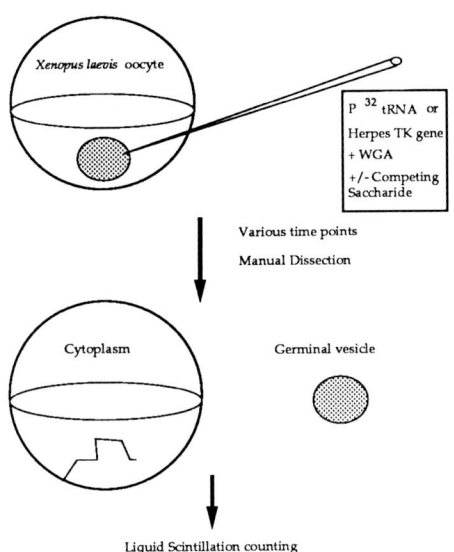

FIGURE 4. RNA transport assay and the effect of WGA.

also inhibited transport. This inhibition was not accompanied by an inhibition of the rate of free diffusion across the pore complex since dextran or inulin diffusion was not altered by the lectin. We interpret these studies to suggest that the lectin interacts with O-linked N-actylglucosamine residues present on some component of the transport machinery-perhaps the nuclear pore itself.

MOLECULAR CLONING OF A NUCLEAR PORE GLYCOPROTEIN

Availability of antibodies to components of the pore complex has allowed the molecular cloning of the major nuclear pore glycoprotein having a molecular mass of 62,000 (26). This protein was purified by immunoaffinity

chromatography and was subjected to microsequencing (Fig 5). In this way, it was possible to obtain two short stretches of protein sequence. Using the appropriate oligonucleotide probe, a cDNA clone which encodes the carboxyl terminal third of the polypeptide was identified. This was used as a probe to isolate a series of overlapping phage clones. From the sequence of the polypeptide predicted to be encoded by these clones, a number of the structural features of p62 protein have emerged. These features are summarized in Fig 6.
The most striking feature of the amino domain of the molecule is the presence of alpha helical stretches with some sequence similarity to the keratins, myosin, actin and tropomyosin. Only a single O-linked GlcNAc residue was found associated with this portion of the molecule; p62 is

Isolation and Sequencing of p62

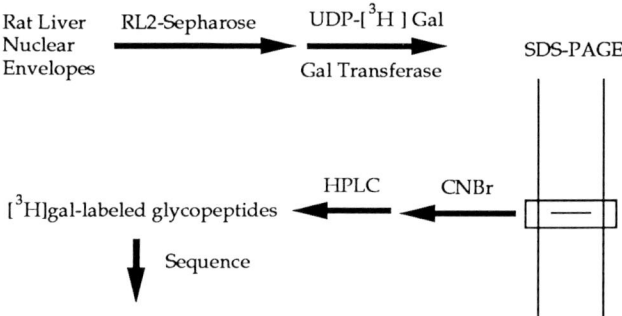

FIGURE 5. Isolation and sequencing of Nuclear pore protein p62.

DOMAIN STRUCTURE OF p62

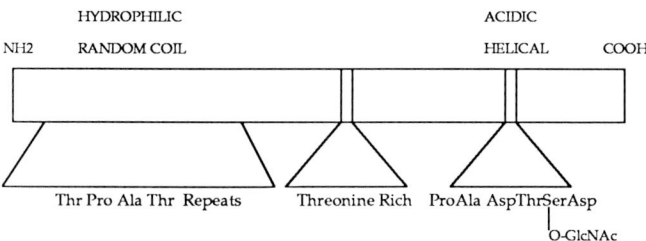

FIGURE 6. Domain structure of p62.

known to be modified at approximately 10 sites (26). The site of glycosyation with O-linked GlcNAc was found to be: Pro Ala Asp Thr Ser-GlcNAc Asp Pro.

Although this is the major protein recognized by pore specific antibodies, the role p62 plays in maintaining the structure of the pore is unknown. When additional structural information becomes available, it should be possible to determine where in the pore complex this portein resides. It may also be possible to use recombinant DNA technology and reverse genetic approaches to correlate the structure of p62 with its possible functions in nucleocytoplasmic exchange.

SUMMARY AND CONCLUSIONS

The morphological and biochemical experiments outlined argue that the nuclear pore is the principal route of nuclear protein import. The problem reduces to the question of how the pore complex accomodates large nuclear proteins and selectively excludes cytoplasmic proteins. It is known that the uptake requires ATP, yet it is unlikely that this energy is involved in the unfolding of nuclear proteins prior to uptake. Studies carried out in our laboratory suggest that when injected into the cytoplasm, antibodies to the SV40 large T antigen are cotransported into the nucleus of SV40 infected cells with the T antigen. These data suggest that the antigen-antibody complex survives the transport event. The uptake of nuclear proteins is quite efficient; our calculations suggest for a protein bearing the SV40 T antigen localization sequence, the uptake rate is greater than 1000 molecules/pore complex/min. Based on the information presented concerning both the structure of the pore complex and the nature of nuclear transport, a number of models for how this process might occur can be envisioned. In all of the models, a receptor is envisioned which specifically recognizes nuclear localization signals present on the surface of nuclear proteins. This receptor would then serve as a means of discrimination nuclear proteins from other proteins in the cytoplasm. The receptor might be envisioned to move along the filaments in an energy dependent manner. Such a facilitated diffusion model would also suggest that a mechanism exists which would promote dissociation of the nuclear proteins from the receptor after transport. This might be accomplished by a post-translational modification of the receptor, such as phosphorylation. In a variant of this hypothesis, occupancy of the receptor by a nuclear protein would trigger pore expansion, allowing selective transport of the nuclear protein-receptor complex into the nucleus. Currently, these and other models are being tested in our laboratory.

REFERENCES

1. Peters, R. (1986). Fluorescence microphotolysis to measure nucleocytoplasmic Transport and intracellular mobility. Biochimica Biophysica Acta 864: 305.

2. Newport, J.W., & Forbes, D.J. (1987). The nucleus: structure, function and Dynamics. Annual Reviews of Biochemistry 56: 535.

3. Unwin, P.N.T., & Milligan, R.A. (1982). A large particle associated with the perimeter of the nuclear pore complex. Journal of Cell Biology 93: 63.

4. Jiang, L.-W., & Schindler, M. (1987). Fluorescence photobleaching analysis of nuclear transport: dynamic evidence for auxiliary channels in detergent-treated nuclei. Biochemistry 26: 1546.

5. Paine, P.L., & Feldherr, C.M. (1972). Nucleocytoplasmic exchange of macromolecules. Experimental Cell Research 74: 81.

6. Feldherr, C.M., Kallenbach, E., & Schultz, N. (1984). Movement of a karyophilic protein through the nuclear pores of oocytes. Journal of Cell Biology 99: 2216.

7. Holt, G.D. & Hart, G.W. (1986). The subcellular distribution of terminal N-acetylglucosamine moieties: Localization of a novel protein-saccharide linkage, O-linked GlcNAc. Journal of Biological Chemistry 261: 8049.

8. Hanover, J.A., Cohen, C.K., Willingham, M.C., & Park, M.K. (1987). O-linked N-acetylglucosamine is attached to proteins of the nuclear pore. Evidence for cytoplasmic and nucleoplasmic glycoproteins. Journal of Biological Chemistry 262: 9887.

9. Hart, G.W., Holt, G.D. & Haltiwanger, R.S. (1988). Nuclear and cytoplasmic glycosylation: Novel saccharide linkages in unexpected places. Trends in Biochemical Sciences 13: 380.

10. Park, M.K., D'Onofrio, M., Willingham, M.C. & Hanover, J.A. (1987). A monoclonal antibody against a family of nuclear pore proteins (nucleoporins): O-linked N-acetylglucosamine is part of the immunodeterminant. Proceedings of the National Academy of Sciences USA 84: 6462.

11. Snow, C.M., Senior, A. & Gerace, L. (1987). Monoclonal antibodies identify a group of nuclear pore complex proteins. Journal of Cell Biology, 104: 1143.

12. Davis, L.I. & Blobel, G. (1986). Identification and characterization of a nuclear pore complex protein. Cell 45: 699.

13. Kalderon, D., Richardson, W.D., Markham, A.F., & Smith, A.E. (1984). Sequence requirements for nuclear location of simian virus 40 large-T antigen. Nature 331: 33.

14. Lanford, R.E., & Butel, J.S. (1984). Construction and characterization of an SV40 mutant defective in nuclear transport of T antigen. Cell 37: 801.

15. Kalderon, D., Roberts, B.L., Richardson, W.D., & Smith, A.E. (1984). A short amino acid sequence able to specify nuclear location. Cell 39: 499.

16. Goldfarb, D.S., Gariepy, J., Schoolnik, G., & Kornberg, R.D. (1986). Synthetic peptides as nuclear localization signals. Nature 332: 641.

17. Lanford, R.E., Kanda, P., & Kennedy, R.C. (1986). Induction of nuclear transport with a synthetic peptide homologous to the SV40 T antigen transport signal. Cell 46: 575.

18. Yoneda, Y., Imamoto-Sonobe, N., Yamaizumi, M., & Uchida, T. (1987). Reversible inhibition of protein import into the nucleus by wheat germ agglutinin injected into cultured cells. Experimental Cell Research 173: 586.

19. Wolff, B., Willingham, M.C., & Hanover, J.A. (1988). Nuclear protein import: specificity for transport across the nuclear pore. Experimental Cell Research, 178, 318-334.

20. Hanover, J. A., D' Onofrio, Starr, C. M. Park, M. K., and Wolff, B. (1988) "Nucleocytoplasmic Transport: Molecular Characterization of the Nuclear Pore." in <u>Advances in Biotechnology of Membrane Ion Transport, p 77.</u> Jorgensen and Verna, eds. Raven Press, New York.

21. Roberts, B.L., Richardson, W.D., & Smith, A.E. (1987). The effect of protein context on nuclear location signal function. <u>Cell</u> 50: 465.

22. Smith, A.E., Kalderon, D., Roberts, B.L., Colledge, W.H., Edge, M., Gillett, P., Markham, A., Paucha, E., & Richardson, W.D. (1985). The nuclear location signal. <u>Proceedings of the Royal Society of London</u>, 226: 43.

23. Dabauvalle, M.-C., Schulz, B., Scheer, U., & Peters, R. (1988). Inhibition of nuclear accumulation of karylphilic proteins in living cells by microinjection of the lectin wheat germ agglutinin. <u>Experimental Cell Research</u> 174: 291.

24. Zasloff, M., Rosenberg, M. & Santos, T. (1982). Impaired nuclear transport of a human variant tRNA^{met}. <u>Nature</u> 300: 81.

25. Zasloff, M. (1983). tRNA-transport from the nucleus in a eukaryotic cell: Carrier-mediated translocation process. <u>Proceedings of the National Academy of Sciences USA</u> 80: 6436.

26. D'Onofrio, M., Starr, C.M., Park, M.K., Holt, G.D., Haltiwanger, R.S., Hart, G.W. & Hanover, J.A. (1988) Partial cDNA sequence encoding a nuclear pore protein modified by O-linked N-acetylglucosamine. <u>Proceedings of the National Academy of Sciences USA</u> 85: 9595.

SELECTION OF AN EXPRESSION HOST FOR HUMAN GLUCOCEREBROSIDASE: IMPORTANCE OF HOST CELL GLYCOSYLATION

Michel L.E. Bergh, Carol Naranjo, Anita F. Mentzer, Gary D. Barsomian, C. William Christopher, Catherine Bartlett, Shirish Hirani and James R. Rasmussen

Genzyme Corporation, Boston, MA 02111

ABSTRACT Human glucocerebrosidase can be expressed in *Spodoptera frugiperda* Sf9 cells by infection with a recombinant baculovirus containing the glucocerebrosidase gene under control of the polyhedrin promoter. The recombinant protein produced by the cells is a glycoprotein, and it is enzymatically active with its physiological substrate. Recombinant glucocerebrosidase is slowly secreted. This is in sharp contrast with the subcellular targeting of human glucocerebrosidase synthesized by human fibroblasts, in which the protein accumulates in lysosomes. During the secretion step the protein undergoes processing of the Asn-linked carbohydrate chains. The processing can be blocked by deoxymannojirimycin, swainsonine, castanospermine, deoxynojirimycin and N-methyldeoxynojirimycin. Except in the case of swainsonine, secretion is also severely impaired. HPLC analysis shows that the reduced carbohydrate chain(s) derived from the secreted protein, co-migrate with $Man_3GlcNAc(Fuc)GlcNAcol$.

INTRODUCTION

Gaucher's disease is an autosomal recessive lysosomal storage disease which is characterized by a deficiency of the lysosomal enzyme, glucocerebrosidase (GCR). In the most common form of the disease (type 1), the substrate for the deficient enzyme, glucocerebroside, accumulates in phagocytic cells, predominantly in the spleen, liver and bone marrow. Clinical manifestations include spenomegaly, hepatomegaly, skeletal disorders, thrombocytopenia and anemia.

It was proposed that Gaucher's disease might be treated by enzyme replacement therapy (1). Most of the work aimed at such a therapy involved the use of GCR purified from human placenta. It was suggested that targeting of placenta-derived GCR to macrophages could be improved by treating GCR with neuraminidase, galactosidase and hexosaminidase, thus generating a mannose-terminating protein (figure 1). In a rat model, it was shown (2) that the glycosidase-treated enzyme was taken up 5-fold more efficiently by Kupffer cells than the native protein, presumably via the Man/GlcNAc receptor. This glycosidase-treated form of placenta GCR is currently in clinical trials.

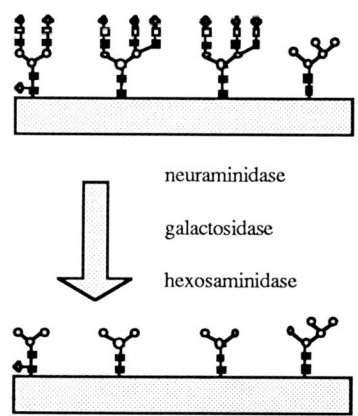

FIGURE 1. Schematic diagram of glycosidase treatment of glucocerebrosidase purified from human placenta, resulting in a mannose-terminated form with increased macrophage-targeting properties.

An alternative to placenta-purified GCR is a recombinant form of GCR produced by host cells expressing the human GCR gene. Recombinant GCR has been expressed in active form in several cell lines, including insect cells (3).

Since carbohydrate-mediated targeting of GCR is an essential aspect of its therapeutic efficacy, the glycosylation pattern of the expression host is an important criterion. It has been reported that the sugar chains of the Sindbis E1 and E2 glycoproteins produced in an infected mosquito cell line (*Aedes albopictus*) consisted only of high-mannose oligosaccharides ranging in size from $Man_9GlcNAc_2$ to $Man_3GlcNAc_2$ (4). These results are in accordance with the observations made previously by Butters and Hughes on the glycosylation in *Aedes aegypti* cells (5) and observations by Nordin et al. (6). Thus, insect cells appear to have glycosylation characteristics which are potentially useful for targeting GCR to its site of action. Moreover, a high-yield expression system, based on recombinant baculovirus infection, has been developed for *Spodoptera frugiperda* cells (7). Therefore, we started to investigate the utility of insect cells for the production of therapeutic GCR. In a first stage, we characterized the glycosylation, processing and secretion of this protein expressed in Sf9 cells.

MATERIALS AND METHODS

Cell Culture and Recombinant Virus Production

The baculovirus *Autographa californica* and the recombinant viruses carrying GCR were used to infect *Spodoptera frugiperda* Sf9 cells at a multiplicity of infection of 10 for expression of GCR and 0.1 for virus production. Sf9 cells were maintained in TNM-FH containing 10% fetal bovine serum.

Construction of the GCR-encoding transfer vectors, co-transfection with viral DNA, subsequent harvesting of recombinant viruses and production of recombinant protein were accomplished by established methods (8). The deletion in the 5' non-coding region of the GCR gene was made by exonuclease III digestion. At the 3' end of the gene, a SacI site at +1844 was used to remove most of the 3' non-coding sequences. The transfer vectors pAc373 and pVL941, the cell line Sf9 and the baculovirus *A. californica* were all generously provided by Dr. M. Summers, Texas A&M University.

Metabolic Labeling

For metabolic labeling, 24 hours p.i. cells were pre-incubated for 30 minutes in methionine-free Grace's medium, after which [^{35}S]methionine was added in a small volume. Generally, labeling was carried out for 15 minutes, after which the labeling medium was removed, the cells were washed and incubated in chase medium (complete TNM-FH). When labeling was done in the presence of inhibitors, these compounds were present during the pre-incubation as well as during the pulse and chase periods.

Purification and Carbohydrate Analysis

Recombinant GCR was purified from the medium by hydrophobic chromatography, essentially as described (9). Carbohydrate analysis of the carbohydrate chains derived from the purified protein was based on N-Glycanase™ digestion, NaB[^{3}H]$_4$ reduction and HPLC analysis of the carbohydrate chain(s) as described by Hirani *et al.* (10).

RESULTS

Vector Constructions

Two vector constructions were made, both of which contain GCR under control of the polyhedrin promoter (figure 2).

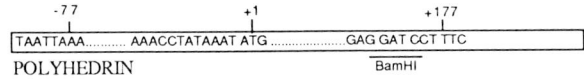
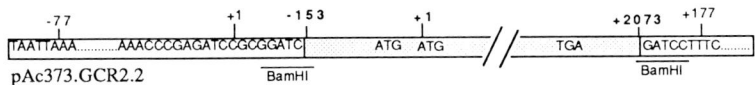
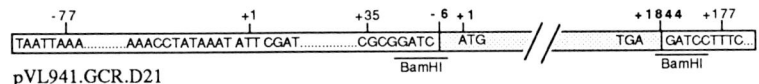

FIGURE 2. Schematic diagram of transfer vectors, showing the position of the GCR gene (shaded bars) in relationship to the regulatory and coding sequences of the polyhedrin gene (open bars). Nucleotide +1 of the polyhedrin gene represents the first nucleotide of the translation start codon. For GCR, nucleotide +1 (in bold print) is the first nucleotide of the second translation start codon.

In pAc373.GCR2.2, most, but not all of the polyhedrin regulatory sequences is present: the construction contains up to nucleotide -8 of the polyhedrin gene, (+1 is the first nucleotide of the polyhedrin start codon). In addition, the construction contains a 2.2 kb cDNA fragment encoding GCR (11), inserted into the BamHI site. This insert contains two potential translation start codons, of which only the second one is preceded by a Kozak consensus sequence. The transfer vector pVL941.GCR.D21 contains the entire 5' non-coding region of the polyhedrin gene, in addition to the first 35 nucleotides of the polyhedrin coding region, with a mutagenized translation start-codon at +1, in order to avoid the production of a fusion protein. The GCR insert in this construction is only 1.8 kb in size. At the 5' end of the gene all nucleotides up to -8 of the second translation start codon have been removed, and at the 3' end the fragment extends only 0.3 kb beyond the translation termination codon.

Expression of Recombinant Glucocerebrosidase

The transfer vectors were co-transfected with wild-type genomic viral DNA. The viral progeny resulting from this procedure was subjected to visual plaque screening. Several of the occlusion-negative viruses (i.e. no polyhedrin expression) were plaque-purified. The majority of these clones encoded full-length GCR, as demonstrated by immunoprecipitation of lysates from metabolically labeled cells infected with recombinant virus (data not shown). With both constructions, a recombinant protein was obtained which was enzymatically active with 4-methylumbelliferylglucoside and the physiological substrate glucocerebroside. Production of rGCR usually starts at 24 h after infection with the recombinant virus, and tapers off after 70-80 hours.

Characterization of Recombinant Glucerebrosidase

In a pulse-chase experiment with [^{35}S]methionine 24 hours post infection it could be demonstrated that the recombinant protein is slowly secreted (figure 3). During the first two hours after the pulse, virtually no radioactive, immunoprecipitable material could be detected in the medium. After approximately 3 hours significant quantities of GCR are released into the medium. The appearance of GCR in the medium is not the result of cell death, since at this timepoint cells are still viable and have not yet reached the stage of virus-induced lysis. The t1/2 of secretion of recombinant GCR was roughly estimated to be approximately 8 hours; after 30 hours 95% of the protein is present in the medium.
The secretion of the protein by Sf9 cells is preceded by a processing step. This step converts the intracellular form, which has an apparent molecular weight of 62 kD, to a form of 58 kD

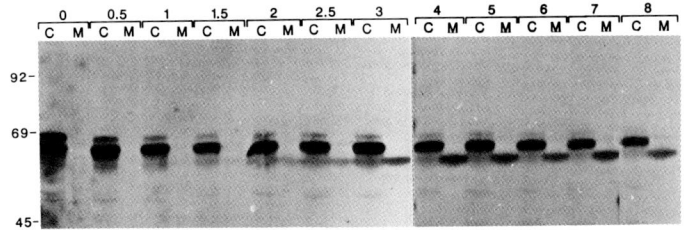

FIGURE 3. Human glucocerebrosidase in the cells and medium of infected Sf9 cells. Infected cells were pulse-labeled (24 h p.i.) for 15 min with [^{35}S]methionine, and then chased for the time-periods indicated (in hours). At the end of the chase period, cells (C) and medium (M) were collected, immunoprecipi-tated with a polyclonal anti-GCR antibody and subjected to SDS-PAGE followed by fluorography.

(figure 3). Because hardly any 58 kD material can be found in the cell lysate, it is assumed that this processing step takes place shortly before the secretion step.

Glycosylation of Recombinant Glucocerebrosidase

The difference in size between intra- and extracellular glucocerebrosidase is due to differences in Asn-linked glycosylation, since both the 62 kD and the 58 kD forms can be converted to a protein of 54 kD by treatment with N-Glycanase™. This is in accordance with the apparent molecular weight of N-Glycanase™-treated GCR purified from human placenta. Furthermore, the intracellular form is susceptible to both endoglucosaminidase H (endo H) and endoglucosaminidase F (endo F), whereas the extracellular form is resistant to both glycosidases (figure 4).

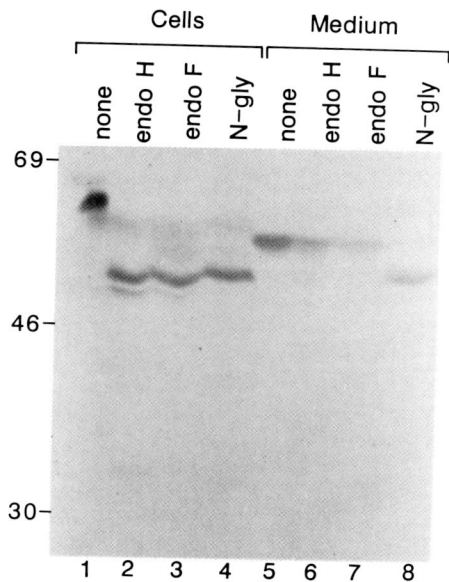

FIGURE 4. Glycosidase treatment of human recombinant GCR. GCR was isolated from medium and cell lysate from [^{35}S]methionine-labeled infected Sf9 cells by immunoprecipitation. Immunoprecipitates were subjected to endo H, endo F and N-Glycanase™ treatments as indicated, and subjected to SDS-PAGE followed by fluorography.

Effects of Oligosaccharide Processing Inhibitors

The processing of the carbohydrate chains of rGCR from endo H sensitive to endo H resistant can be blocked by various processing inhibitors.
Two mannosidase inhibitors were tested, swainsonine and deoxymannojirimycin. In a pulse-chase study, swainsonine appeared to block carbohydrate processing of rGCR completely, without affecting secretion to any significant extent (figure 5, upper panel). Deoxymannojirimycin, on the other hand, does not appear to inhibit processing at concentrations which have been shown

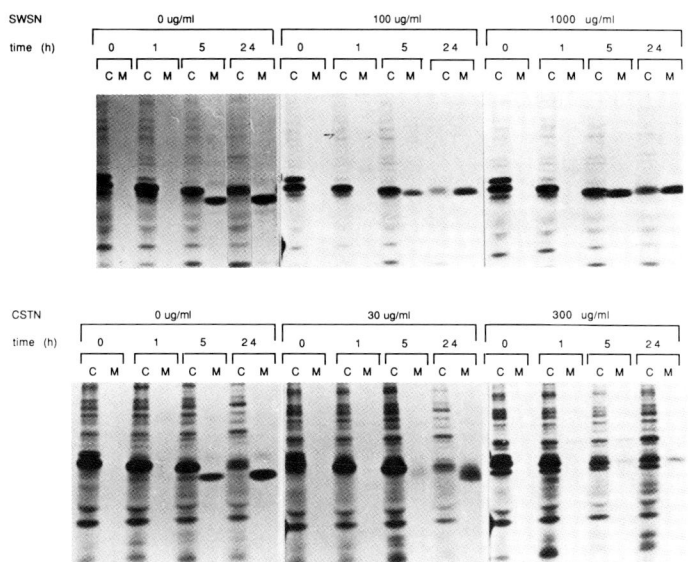

FIGURE 5 Effect of swainsonine (top) and castanospermine (bottom) on processing and secretion of rGCR. Cells were pulse-labeled 24 hours after infection with recombinant virus and chased for the times indicated. Medium (M) and cells (C) were collected and subjected to immunoprecipitation and SDS-PAGE. The inhibitors were present at the indicated concentrations throughout the pulse and chase periods, as well as during a 1 hour pre-incubation.

to be effective in mammalian cells (1 mM). Only at very high concentrations (10 mM) can processing be blocked by this inhibitor. At these concentrations, however, secretion is virtually abolished (data not shown).

The glucosidase inhibitor castanospermine has a similarly deleterious effect on GCR secretion. At concentrations high enough to prevent oligosaccharide processing, the majority of the radioactive immunoprecipitable material can be found associated with the cells, even after 24 hours (figure 5, lower panel). The material appearing in

the medium has a molecular weight of approximately
62 kD, but the GCR associated with the cells seems
to undergo some form of processing. It is not
known if this processing is a proteolytic step or
hydrolysis of the oligosaccharide moiety by
castanospermine-insensitive glycosidases. Two
other glucosidase inhibitors, N-methyldeoxynojirimycin and deoxynojirimycin, had very similar
effects (data not shown).

Oligosaccharide Structural Characterization

Recombinant GCR was purified from the con-
ditioned medium of cells infected with

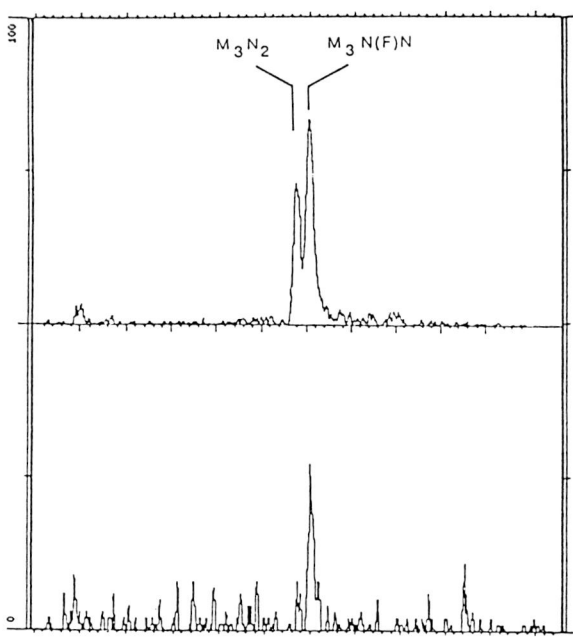

FIGURE 6. HPLC chromatography of reduced
oligosaccharides on Micropak AX-5. Oligosaccharides were isolated by N-glycanase™ treatment,
reduced with sodium borotritide, and analysed by
HPLC. Upper panel: oligosaccharide fraction from
glycosidase-treated placenta GCR (see figure 1).
Lower panel: oligosaccharide fraction released
from Sf9-expressed secreted GCR.

Ac373.GCR2.2. The Asn-linked carbohydrate chains were released by exhaustive N-Glycanase™ digestion and analyzed as described (10). As a control, a similar analysis was carried out using the glycosidase-treated form of human placenta-derived GCR. The latter protein contains two different carbohydrate structures, $Man_3GlcNAc_2$ and $Man_3GlcNAc(Fuc)GlcNAc$, respectively (figure 6, upper panel). The oligosaccharide fraction obtained from the secreted form of Sf9-derived rGCR appears to be relatively homogenous, since only one peak could be detected on HPLC, which seems to co-migrate with the $Man_3GlcNAc(Fuc)GlcNAc$ fraction of the placenta-derived GCR (figure 6, lower panel).

Although the possibility cannot yet be entirely excluded, there is no evidence so far for the presence of O-linked carbohydrate chains on Sf9-derived GCR.

DISCUSSION

The vast majority of recombinant proteins which so far have been produced in insect cells using the baculovirus expression system, are either cytosolic, cell-surface or secretory proteins or glycoproteins (for a review see ref. 12). Human GCR is, to our knowledge, the first mammalian lysosomal protein expressed in *S. frugiperda* cells. Lysosomal targeting in mammalian cells can be mediated by the mannose-6-phosphate receptor system as well as via a carbohydrate-independent pathway. It appears that, at least in human skin fibroblasts, GCR is targeted to lysosomes via the second route (13). It is not known if insect cells also contain these two lysosome targeting pathways, and if they exist, how homologous these are with their mammalian counterparts. We and others (3) observe that human GCR is secreted by Sf9 cells, instead of accumulating intracellularly. This suggests that the lysosomal targeting pathways of Sf9 cells, Man-6-P mediated or otherwise, fail to interact with human GCR. On the other hand, it is not clear why, in the absence of any ER- or Golgi-localized recognition system, the secretion of GCR is so extremely slow.

By comparison, the half-life of secretion of human tPA in Sf9 cells is 1.6 h (14), whereas that of GCR is roughly 8 h. A highly speculative hypothesis may be that some sort of (possibly Man-6-P-independent) recognition system does exist in Sf9 cells which interacts with GCR. However, due to poor recognition of the heterologous protein, this systems fails to target GCR properly to lysosomal compartments, but has enough affinity to slow down its secretion.

The processing of the carbohydrate chains of GCR in Sf9 cells is somewhat similar to the situation observed in mammalian cells. At first, the protein is endo H sensitive, indicating that it contains carbohydrate chains of the size of $Man_4GlcNAc_2$ or larger. In a later stage, the protein becomes endo H resistant. In contrast to mammalian glycosylation, though, in Sf9 cells the carbohydrate chains are processed to smaller species. Using HPLC analysis, we could show that in the case of GCR, these appear to have the same size as $Man_3GlcNAc(Fuc)GlcNAc$. This constitutes the first direct biochemical analysis of an oligosaccharide fraction derived from an Sf9-expressed glycoprotein. Although the two carbohydrate fractions (Sf9- and placenta-derived, respectively) have the same apparent size on HPLC, this does not imply an identical structure. Nevertheless, our result certainly seems encouraging enough to justify further investigation of the therapeutic utility of Sf9-derived GCR.

The observation that the processing step from endo H sensitivity to endo H resistance takes place just before the protein is released into the medium, may be another argument against a lysosomal localization of intracellular GCR. In addition, the processing of the oligosaccharides of GCR could be blocked by processing inhibitors which are known to inhibit late ER/cis Golgi processing enzymes. It seems therefore more likely that the majority of the intracellular GCR is present in the ER, and that the rate-limiting step is transport through the ER. As soon as the protein reaches the compartments where processing takes place, secretion follows rapidly thereafter.

It should also be pointed out that cleavage of the signal sequence of human rGCR seems to be carried out efficiently by Sf9 cells. This can be concluded from the observation that the size of the mature, deglycosylated Sf9-expressed GCR (54 kD) is very similar to the size of the mature, deglycosylated protein purified from placenta.

Finally, the secreted protein is enzymatically active with its physiological substrate, glucocerebroside, and the specific activity of the purified protein appears to be similar to that of the placenta-derived GCR. Taken together with the fact that the protein is glycosylated, and possibly contains mannose-terminated carbohydrate chains, this suggests that Sf9-derived GCR is a potential candidate for clinical application. Therefore, we are now carrying out a more detailed analysis of the oligosaccharide structures and are investigating the utility of the these glycans in targeting the protein to its therapeutic site of action, the lysosomes of phagocytic cells.

REFERENCES

1. Brady RO (1966). The sphingolipidoses. N Engl J Med 275: 312.
2. Furbish FS, Steer CJ, Krett NL, Barranger JA (1981). Uptake and distribution of placental glucocerebrosidase in rat hepatic cells and effects of sequential deglycosylation. Biochim Biophys Acta 673: 425.
3. Martin BM, Tsuji S, LaMarca ME, Maysak K, Eliason W, Ginns EI (1988). Glycosylation and processing of high levels of active human glucocerebrosidase in invertebrate cells using a baculovirus expression vector. DNA 7: 99.
4. Hsieh P, Robbins PW (1984). Regulation of asparagine-linked oligosaccharide processing: oligosaccharide processing in *Aedes albopictus* mosquito cells. J Biol Chem 259: 2375.
5. Butters TD, Hughes RC (1981). Isolation and characterization of mosquito cell membrane glycoproteins. Biochim Biophys Acta 640: 655.
6. Nordin JH, Gochoco CH, Wojchowski DM, Kunkel JG (1984). A comparative study of the size-

heterogeneous high-mannose oligosaccharides of some insect vitellins. Comp. Biochem. Physiol. 79B: 379.
7. Smith GE, Summers MD, Fraser MJ (1983). Production of human beta interferon in insect cells infected with a baculovirus expression vector. Mol Cell Biol 3: 2156.
8. Summers MD, Smith GE (1987). A manual of methods for baculovirus vectors and insect cell culture procedures. Texas Agricultural Experiment Station Bulletin no. 1555.
9. Murray GJ, Youle RJ, Gandy SE, Zirzow, GC, Barranger JA (1985). Purification of beta-glucocerebrosidase by preparative-scale high-performace liquid chromatography: the use of ethylene glycol-containing buffers for chromatography of hydrophobic glycoprotein enzymes. Anal Biochem 147: 301.
10. Hirani S, Bernasconi RJ, Rasmussen RJ (1987). Use of N-Glycanase to release asparagine-linked oligosaccharides for structural analysis. Anal Biochem 162: 485.
11. Sorge J, West C, Westwood B, Beutler E (1985). Molecular cloning and nucleotide sequence of human glucocerebrosidase. Proc Natl Acad Sci USA 82: 7289.
12. Luckow VA, Summers MD (1988). Trends in the development of baculovirus expression vectors. Bio/technology 6: 47.
13. Aerts JMFG, Schram AW, Strijland A, van Weely S, Jonsson LMV, Tager JM, Sorrell SH, Ginns EI, Barranger JA, Murray GJ (1988) Glucocerebrosidase, a lysosomal enzyme that does not undergo oligosaccharide phosphorylation. Biochim Biophys Acta 964: 303.
14. Jarvis DL, Summers MD (1989). Glycosylation an secretion of human tissue plasminogen activator in recombinant baculovirus-infected insect cells. Mol Cell Biol 9: 214.

GLYCOCONJUGATES AS DRUGS

Russell Greig and George Poste

Smith Kline and French Laboratories
King of Prussia, PA 19406

ABSTRACT The recognition that oligosaccharide structures can influence the pharmacological performance of recombinant macromolecules including tissue plasminogen activator and erythropoietin has elicited significant interest in the contribution of carbohydrate chemistry to the drug discovery process. This interest has been intensified by rapid progress in both analytical and synthetic chemistry which now permits the unravelling of detailed structure-activity relationships for polypeptides bearing simple- and complex-type oligosaccharide side chains. Current attention on this class of (macro)molecule overlooks, however, the immense past and present contribution of carbohydrates and carbohydrate-containing structures to the treatment of human disease. Several low molecular weight glycoconjugates have had, and in many cases continue to have, a profound impact on the management of several disorders including congestive heart failure (digoxin), anticoagulant therapy (heparin), cancer (adriamycin, etoposide) and infectious diseases (streptomycin, erythromycin). In this article the history of these agents is reviewed along with the major features of their pharmacology. While in most cases the role of the carbohydrate structure in influencing their mechanism of action remains obscure, the fact that all these drugs share some degree of carbohydrate character demonstrates clearly the importance of this functionality in the pharmacological performance of a wide range of exceptionally effective drugs.

INTRODUCTION

The importance of glycoconjugates as pharmaceutical agents for the treatment of human diseases has gained increased recognition over the last few years due mainly to the success of biotechnology in providing macromolecules (biopharmaceuticals) for evaluation in a number of clinical disorders. Examples include tissue plasminogen activator (myocardial infarction), the interferons (infectious diseases and certain cancers), colony stimulating factors (myelosuppression), erythropoietin (anemia associated with renal insufficiency), and more recently a soluble version of the human T-lymphocyte CD4 receptor recognized by and responsible for HIV infectivity (AIDS), to name just a few. All these agents are polypeptides and have been produced for clinical trials by cloning and expressing the appropriate gene into either prokaryotes (usually E. coli), yeast or mammalian cells. Final choice of expression system is influenced by several important considerations, including level of expression, ease of purification, stability and specific activity of the final product, and cost. Regardless of the system selected, however, an issue of increasing concern is the glycosylation pattern of the expressed protein and if (and how) it differs from the natural product (1). These concerns are real since the glycosylation machinery of E. coli and yeast are distinct, and differ considerably from (human) mammalian cells. Moreover, the glycosylation patterns enforced upon a heterologous gene product by a particular (mammalian) cell type may reflect the species from which the cell was derived and also its histological origins. Thus if a specific macromolecule is normally synthesized by human macrophages but for technical convenience is cloned, expressed and purified in a hamster epithelial cell or yeast, then there can be little guarantee that the glycosylation profiles of the recombinant and natural molecules will be identical. In fact, the opposite might be anticipated. This is especially true for E. coli which lacks the capacity to glycosylate heterologous macromolecules and synthesizes recombinant proteins stripped of their oligosaccharide structures.

Given these concerns a critical question emerges; how important is carbohydrate structure in determining the pharmacological performance of recombinant macromolecules when administered to man? A few years ago this question was unanswerable, but with well-documented advances in biotechnology and significant progress in analytical and structural carbohydrate chemistry, we can map precisely the glycosylation patterns of natural and recombinant molecules, and modify selectively the carbohydrate structure of a particular macromolecule with subsequent interrogation of its in vitro and in vivo activity. In other words we now possess the technological skills to establish detailed structure-activity relationships for high molecular weight glycoproteins. Apart from its intrinsic scientific value in establishing the role of carbohydrates in the folding and biochemical properties of glycosylated macromolecules, this exercise will likely have significant impact in the design of more effective biopharmaceuticals. One topical example is tissue plasminogen activator (Activase®, Genentech) (2). Activase® is administered intravenously to myocardial infarction patients and exerts its action by converting plasminogen to the serum protease plasmin, which in turn dissolves the offending thrombus thus re-establishing blood flow through the coronary vessel. Large quantities (100 mg) of Activase® are required to achieve clot dissolution, and its plasma half-life is short (approximately 4 min). The factors influencing the rapid clearance of Activase® from the circulation remain unclear, but there is some evidence that the oligosaccharide structure may play an important role (2). A logical strategy then is to modify the glycosylation pattern of tissue plasminogen activator with the objective of reducing clearance rate and extending plasma half-life. If this were achieved, less drug would be required to elicit a therapeutic effect. Given the high cost of Activase® treatment (ranging from $2100 to $4500 per 100 mg) and since it is the only recombinant tissue plasminogen activator currently approved by FDA, any maneuver to reduce dosage (and thus price) while preserving (or improving) therapeutic efficacy could readily generate a competitive product. Issues like this along with concerns on the antigenicity of recombinant molecules whose oligosaccharide structure differs from the natural molecule (3,4), have highlighted the importance of carbohydrate chemistry in developing new biopharmaceutical products and improving upon existing ones, and this topic is now the center of considerable academic and industrial attention.

Nevertheless the current debate centered on biopharmaceuticals frequently overlooks the fact that carbohydrate structures have already contributed immeasurably to the discovery of important drugs and the treatment of human disease. The agents in question (ranging from digoxin to streptomycin) are not macromolecules but represent lower molecular weight agents decorated with mono-, di-, and (less frequently) trisaccharides. With few exceptions, all have been isolated from natural sources (microbial fermentation broths, plants and marine extracts). The purpose of this article is to describe the most important of these structures, and in doing so illustrate the past contributions of glycochemistry to drug discovery.

Relatively small glycoconjugate structures feature prominently in the treatment of several classes of human diseases including cardiovascular conditions, cancer and infectious diseases, and examples of each will be considered in turn.

Cardiovascular

Digoxin The identification of digoxin for the treatment of heart failure is an instructive example of drug discovery through a combination of folklore medicine and thorough biochemical detective work (5). The medicinal properties of the foxglove plant (_Digitalis purpurea_) have been traced back to the 14th century, but it was not until the 1770s that its pharmacological utility in patients suffering from dropsy (generalized edema, a sequelae to congestive heart failure) was carefully interrogated (6,7). Over a nine year period an English physician, William Withering, treated over 150 patients with digitalis (powdered foxglove leaves) and summarized his results in 1785 in an article entitled "An account of the Foxglove and Some of its Medical Uses; with Practical Remarks on Dropsy, and Other Diseases"(8). Two thirds of his patients responded, thus establishing digitalis as an effective treatment for dropsy, although it was almost another century before digitalis was recognized as exerting its effect through cardiac stimulation (9-11). Starting in the 1820s several attempts were made to isolate the active principle from _Digitalis_ _purpurea_, culminating in the unravelling of the digitoxin structure in 1928 by Adolph Windaus, a considerable feat given state-of-the-art carbohydrate chemistry in the earlier part of this century. Meanwhile it had also been discovered that a related plant, _Digitalis_ _lanata_ (woolley foxglove), displayed greater activity than digitoxin (5), and in the late 1920s through the efforts of Sydney Smith (at Burroughs Wellcome, UK) this novel component was identified as digoxin. Because it binds less tightly to plasma proteins and has a shorter half-life, digoxin is now favored over other glycoside cardiostimulants, including digitoxin and oubain. With the availability of pure material, the clinical pharmacology of digoxin has been investigated in detail and even now over fifty years since its chemical identification, digoxin remains front line therapy for the treatment of congestive heart failure.

Digoxin

Like other cardiac glycosidases, digoxin is composed of an aglycone ring conjugated to an oligosaccharide structure composed of one (ouabain), two (digitalin) or three (digoxin) sugar residues. The pharmacophore or "business end" of the molecule resides in the aglycone structure, while the carbohydrate moiety appears to play a critical role in influencing pharmacokinetics (12).

Digoxin is currently FDA-approved for congestive heart failure, atrial fibrillation, atrial flutter and paroxysmal atrial tachycardia. Biochemically it operates through direct and indirect mechanisms. The direct mechanism involves inhibition of the Na^+/K^+ potassium ATPase pump in cardiac muscle leading to an increase in calcium flux and enhanced myocardial contractility. The indirect action is mediated by the autonomic nervous system and, through a vagomimetic action, is responsible for digoxin's effects on the sinoatrial and atrialventricular nodes. Pharmacologically the net outcome of these actions is that digoxin acts as a positive ionotrope but, and this is important, has the added advantage of slowing heart rate (bradycardia). The latter effect is of particular benefit in patients with heart disease, and it is this combination of ionotropicbradycardiac action that distinguishes digoxin from other sympathomimetic ionotropes (which tend to elicit tachycardia) and encourage its continual use

(12). However, this agent is not without its problems, the major one being the narrow concentration range separating its therapeutic and toxic effects. The latter, which include intense vomiting, has received some unusual documentation in that Leonhard Fuches, the botanist who coined the term <u>Digitalis purpurea</u> in 1542, described digitalis as a "violent medicine", while the poisonous properties of the foxglove plant earned it the Scottish nickname of "deadman's bells". In addition, ouabain, a cardiac glycoside with a single saccharide unit was originally isolated from the ouabaio tree, and provided the Somalis of East Africa with their source of arrow poison (5).

Ouabain

Efforts at identifying safer digoxin analogs have not been successful (13,14), reflecting the fact that the toxicities of this class of agents are, for the most part, an extension of their pharmacology. Few studies have embarked upon methodical alterations to the oligosaccharide component due primarily to the inherent technical complexities of this task. Consequently it is worth pondering whether the sophisticated carbohydrate chemistry of the late 1980s may provide, 50 years after the discovery of the parent compound (15), a more effective digoxin. This retrospective application of carbohydrate chemistry to the improvement of low molecular weight glycoconjugate drugs is a question that can be posed for all the structures described in this article.

<u>Heparin</u> Until the beginning of this century the clotting of blood was a insurmountable obstacle in performing donor to patient transfusions. Although some success was encountered with hirudin, the anticoagulant isolated from leeches, major progress was not made until the 1920s when two groups working independently in France and at John Hopkins University discovered an anticoagulant activity in extracts of dog liver (5,16,17). This material was termed heparin and was identified as a

sulphur-containing polysaccharide. Following this observation other workers discovered a distinct but related heparin in extracts of beef lung. Purified material, confirmed as an acidic sulfated polysaccharide, was clinically evaluated for its ability to block thrombosis in traumatized or surgically treated patients. The technique was successful and led to the rapid acceptance of heparin as front-line anticoagulant therapy. It proved particularly useful in World War II when it saved many lives by permitting battlefield donor to patient transfusion.

Heparin is now known to be a heterogenous group of straight-chained polysaccharides of varying length, whose precise structure is variable.

(1) a-L-iduronic acid 2-sulfate
(2) 2-deoxy-2-sulfamino-a-D-glucose 6-sulfate
(3) B-D-glucuronic acid
(4) 2-acetamido-2-deoxy-a-D-glucose
(5) a-L-iduronic acid

Present in decreasing amounts, usually: (2)>(1)>(4)>(3)>(5).

Heparin

Heparin is currently approved by FDA as an anticoagulant to prevent thrombosis and embolism and as an adjuvant to transfusions, extracorpeal circulation and dialysis. Its mechanism of action remains descriptive. Heparin blocks clotting by disrupting (in concert with antithrombin III) several critical steps in the coagulation cascade including inactivation of Factor X and thrombin, antagonism of prothrombin conversion to thrombin, and by inhibiting the activation of fibrin-stabilizing factor. Exactly how this is accomplished biochemically is not clear, but involves binding of the polyanionic heparin molecule to proteins of the coagulation reaction, rendering them unfit to participate in the clotting mechanism.

Because of uncertainties in its mechanism of action and inherent difficulties in synthesizing polysaccharide analogs of this complexity, efforts at discovering heparins with improved therapeutic utility have not been rewarding. However, this might be changing. Recently two new low molecular weight fragments of heparin (Fragmin® and Fraxiparine®) have been approved in certain European countries for prevention of thrombosis in patients undergoing surgery or dialysis (18,19). Fragmin® has a longer half-life than heparin which may make for more convenient administration. Several other low molecular weight heparin fragments are expected to be launched in Europe over the next two years. Whether these agents will confer significant benefits compared to the parent molecule has yet to be determined.

Antineoplastics

Doxorubicin Also known as Adriamycin®, doxorubicin, medically and commercially, is the most successful antineoplastic drug yet identified. This compound, discovered as a fermentation product of Streptomyces peucetius by scientists at Farmitalia in 1967, belongs to the anthracycline class of compounds and consists of a water-soluble aminosugar (daunosamine) conjugated through a glycosidic bond to a water-insoluble tetracyclic component, adriamycinone (20).

Doxorubicin

Doxorubicin has FDA approval for the treatment of a number of cancers including acute myelocytic leukemia, acute lymphocytic leukemia, Hodgkins and non-Hodgkins lymphoma, Wilms's tumor, neuroblastoma, sarcomas, breast, ovarian and transitional cell bladder carcinoma, and bronchiogenic carcinoma. Its mechanism of action is unclear and controversial. It is well established that doxorubicin, like other anthracyclines, intercalates with DNA thus blocking synthesis of nucleic acids, but whether this reaction in itself can account for all of its antineoplastic properties is not known. Doxorubicin is a reactive molecule and can perturb a number of biochemical processes. The contribution of the sugar group to pharmacological activity appears to be threefold. It provides hydrophilic character to enhance solubility, it may impart a degree of chemical stability to the conjugate, and it appears necessary for the tight intercalation between the anthracyline moiety and adjoining nucleotide bases in DNA (20).

In keeping with most other antineoplastic drugs, doxorubicin has several toxic liabilities including bone marrow suppression (acute) and cardiotoxicity (cumulative). Despite efforts to identify analogs lacking these undesirable traits more effective structures have been proven elusive (21,22). Again, it is interesting to speculate whether systematic analoging around the amino sugar using state-of-the-art carbohydrate chemistry may provide a partial solution to this clinically important problem.

Bleomycin Bleomycin is also a natural product and was originally isolated from fermentation broths of Streptomyces verticillus in 1962 at the Institute of Microbial Chemistry in Japan (20,23). The antitumor activity was attributed to a complex mixture of structurally related glycopeptides, the most prominent of which is bleomycin A_2. These agents are sulfated polypeptides attached to a series of sugar residues.

Bleomycin A_2

Bleomycin has FDA approval for the palliation of several types of cancer including squamous cell carcinoma, lymphomas and testicular cancer. The compound appears to act by binding to DNA and causing both single- and double-stranded scission, and as a result blocks DNA synthesis. Bleomycin lacks myelosuppressive activity but does cause significant lung damage (fibrosis), its major clinical liability.

Because of its complexity, structure-activity relationships for bleomycin have been restricted mainly to a comparative analysis of naturally occuring analogs, although a total organic synthesis of the parent molecule has been achieved (24). The role of the saccharide component in influencing solubility, stability, pharmacokinetics and pharmacological performance remains obscure. However there is some evidence that the aglycone structure alone can bind to and induce DNA scission in cell free systems, but not in intact cells, suggesting that the carbohydrate moiety may play some role in membrane transport (23,25).

Once again these speculations must be viewed with caution because of uncertainties in defining the precise reasons for bleomycin's antineoplastic activity, and the debatable relevance of the in vitro models used for such analysis (eg. DNA binding) compared to conditions encountered in the clinic. Nonetheless an interesting question to pose is whether the lung toxicity of bleomycin might be mitigated by methodical alteration of the carbohydrate component.

Mithramycin This agent also known as mithracin and aureolic acid, was first isolated from Streptomyces argillaceus at Abbott Laboratories in the early 1960s (20,26,27). Mithramycin consists of a polycyclic chromophoric structure linked to two oligosaccharide chains.

Mithramycin

This agent is a potent cytotoxic in vitro but displays only moderate antineoplastic activity in animal tumor models (28). Clinically it is most useful against disseminated embryonal cell carcinoma of the testes, but because of its considerable toxicity in vivo mithramycin is now restricted to the management of hypercalcemia associated with malignant disease. The antitumor effect, albeit limited, probably reflects binding of mithramycin to DNA with subsequent inhibition of nucleic acid synthesis, however, the mechanism of action underlying its calcium-lowering property is not understood . Some evidence suggests that mithramycin disrupts osteoclast function and acts as an antagonist of parathyroid hormone but the data are weak (29). Very few structure-activity studies have been conducted on mithramycin, and the importance of the "winged" oligosaccharide structures has not been established.

Etoposide Etoposide, also known as VP-16, Vepesid and epipodophyllotoxin, is an excellent example of a natural product whose antineoplastic activity was enhanced through synthetic modification (5,20,30,31). The parent molecule, podophyllotoxin, was originally isolated from the May apple plant (or American mandrake) and found wide acceptance in the mid 19th century as a purgative. However, it was later recognized that this agent was also a mitotic poison but failed to find clinical applications because of its severe and unacceptable toxicity (5). Efforts to minimize this liability while preserving its antineoplastic properties led Sandoz Laboratories in the early 1970s to prepare several phodophyllotoxin analogs, with the eventual identification of etoposide. Note that the major difference between this compound and the parent molecule is the introduction of carbohydrate character.

Podophyllotoxin   Etoposide

FDA has approved etoposide for the treatment of testicular tumors and small cell lung cancer. Unlike most antineoplastic drugs etoposide is orally active although this route of administration is approved only for the lung indication. Mechanism of action is unclear but appears to involve binding to and scission of (single-stranded) DNA (30). While addition of the saccharide structure clearly endows the parent phodophyllotoxin molecule with clinically useful antineoplastic activity, the reasons(s) underlying this favorable shift in pharmacological performance are not understood, and thus additional efforts aimed at further improvement in therapeutic efficacy must proceed without a rational basis.

Antiinfectives

There is perhaps no better example of the impact of low molecular weight glycoconjugates on human disease than the treatment of bacterial infections by antibiotics. Many of these agents, too numerous to be listed here, contain mono- or disaccharide structures, and below a few select examples are described.

Streptomycin Through the pioneering work of Selman Waksman (5,32-34), who was interested in identifying antibiotics active against penicillan-resistant bacteria, in particular tuberculosis (caused by Mycobacterium tuberculosis), an agent was discovered in 1943 in the fermentation broth of Streptomyces griseus that appeared to display the desired activity (35). This material, termed streptomycin, was the first aminoglycoside to be marketed and was found to be active against a number of infectious diseases refractory to penicillin, including tuberculosis, plague (Pasturella pestis), brucellosis and certain bacterial dysenteries.

Streptomycin

Streptomycin exerts its bactericidal actions by binding irreversibly to "receptor" proteins on the 30 S ribosomal subunits, disrupting mRNA binding and inhibiting bacterial protein synthesis. However, this agent lacks the safety profile of penicillin, and causes neurotoxicity resulting in deafness (34). Apart from its dramatic entrance as an effective treatment for tuberculosis, the discovery of streptomycin is also notable for setting the precedent that galvanized the post-World War II pharmaceutical industry into undertaking an expansive natural screening program that would lead, in turn, to the discovery of other equally important antibiotics, including erythromycin (5).

Erythromycin   As soil samples from around the globe were screened by pharmaceutical companies for bactericidal activity, scientists at Eli Lily isolated in 1952 an agent from Streptomyces erythreus (from a soil sample in the Phillipines) which displayed a range of activities similar to penicillin (5). It was called erythromycin.

Erythromycin

Because erythromycin is structurally distinct from penicillin it has significant clinical value in treating patients allergic to penicillin or with penicillin-resistant staphylococcal infections. It works by inhibiting bacterial protein synthesis (36).

Amphotericin   Amphotericin, better known as Fungizone®, belongs to the polyene class of antibiotics. It was first isolated in 1953 by the Squibb Institute for Medical Research from a fermentation broth of Streptomyces nodosus, originally found in a soil sample from Venezuela (5,37,38).

Amphotericin

Like other polyenes (eg. nystatin), amphotericin is a highly toxic molecule, and because of its poor absorption can only be used intravenously. Amphotericin is FDA-approved for progressive, potentially fatal infections, particularly those that are systemic. Infections of this type are commonly found in immunocompromised populations such as cancer patients undergoing chemotherapy and increasingly in patients suffering from AIDS. In the latter case, amphotericin is especially useful in the attempted management of disseminated fungal infections. Its mechanism of action is unclear but most likely involves binding to membrane sterols with concomitant breach of membrane integrity.

During the 1950s when these three agents (streptomycin, erythromycin and amphotericin) were discovered, analytical and synthetic carbohydrate chemistry were not as prominent as they are today nor was this type of expertise widely disseminated. As a result, attempts to construct informative structure-activity relationships for these and related compounds were not possible using conventional synthetic chemistry but relied instead on structural analogs identified in the same or similar fermentation broths. This has given us a rather incomplete picture of the role of the carbohydrate residues in influencing the pharmacological properties and mechanism of action of these highly effective antibiotics.

Preclinical Development   The drugs described above are all well established compounds that have been on the market for several years. However, several newer glycoconjugates are also at an earlier stage of development, and serve to illustrate the continual importance of carbohydrate structures as pharmacologically active agents. One example is the class of antitumor antibiotics known as elsamicins. These agents were discovered by the Bristol-Myers Research Institute in Japan in 1986 as products of an actinomycete found in El

Salvador. They were originally detected by their antibacterial and antineoplastic activity, and chemical resolution revealed two related structures, differing only in their carbohydrate composition (39).

Elsamycin A   Elsamycin B

Elsamycin A, which contains an amino sugar in its disaccharide structure, is much more soluble than elsamycin B, and displays greater bactericidal activity, and far more potent antitumor properties. The more impressive pharmacology of elsamycin B is almost certainly attributable to the aminosugar imparting greater solubility, although the mechanism of action of elsamycin B is still unclear.

Another interesting group of compounds currently receiving significant attention are the indole carbazole antibiotics, the most prominent of which is staurosporine, a highly cytotoxic molecule. This compound, isolated from an actinomycete, was first recognized as an antifungal agent but was later shown to be a potent inhibitor of phospholipid/calcium-dependent protein kinase (protein kinase C), with an IC_{50} of 2.7 nM (40).

Staurosporine

Inhibition is not specific, however, since staurosporine also antagonizes with approximately equal potency cAMP-dependent protein kinase from bovine heart (40,41).

Recently several new members of this class of compound have been isolated from fermentation cultures, and differ only in their carbohydrate components (42,43). They have the following structures:

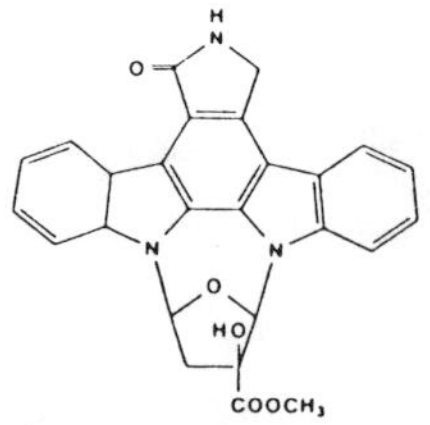

K252a 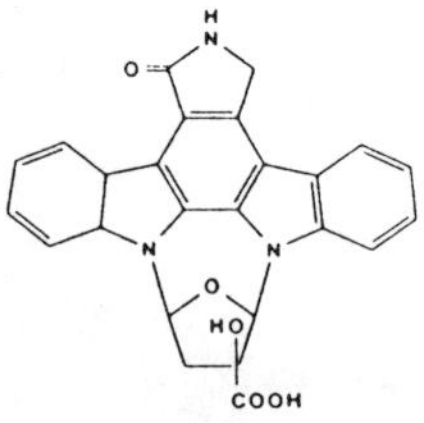   K252b 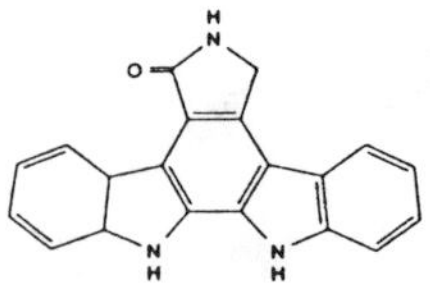   K252c

A comparative analysis of their relative potency in vitro reveals that K252c (the aglycone structure) is less efficacious by an order of magnitude in inhibiting both protein kinase C and phosphodiesterase. K252a and K252b are essentially equipotent. No data have been presented yet on their activity against cAMP-dependent protein kinases.

Comparative Activities of K252 Series of Antibiotics

Compound	R	IC_{50} (nM)	
		PK-C	PDE
K252a	CH_3	33	3
K252b	H	38	11
K252c	aglycone	214	297

The pharmacology of these agents is poorly defined. Hypotensive properties have been associated with staurosporine while the K252 series is claimed to have "serious" effects on platelets, mast cells and vascular smooth muscle (43). Nevertheless, these agents, particularly staurosporine, have served as useful tool compounds in probing protein kinase function in vitro, and given the evidence for an important role for protein kinase C in a number of pathphysiological processes, it is likely that the indole carbazole class of agents (or related compounds) may emerge as clinically useful agents.

Conclusions The drugs and compounds described in this article demonstrate that low molecular weight glycoconjugates have had and continue to have considerable impact on the management of human disease, with their spectrum of therapeutic utility ranging from the spectacular (streptomycin, erythromycin) to the effective (digoxin) and the palliative (bleomycin). These compounds share nothing in common, of course, except their source (natural products) and their differing degrees of carbohydrate character. Consequently, attempts at establishing conclusions on the role of carbohydrate structures in influencing pharmacological performance for any single compound, let alone an entire class, is very difficult, for two main reasons. First, in every case limitations in the scope of synthetic and analytical carbohydrate chemistry has restricted the establishment of structure-activity relationships for the saccharide component, thus clouding any interpretation of their functional importance. Second, the mechanism of action for several of these drugs has proven elusive, and without this insight it is frustratingly difficult to equate structural modifications to the parent compound with specific alterations in pharmacological performance. For example, if the theoretical addition of a disaccharide to an novel aglycone structure results in a more potent compound this could be due to any number (or combination) of reasons including but not limited to solubility, chemical and metabolic stability, absorption, pharmacokinetics, membrane transport and interaction with the final target (e.g. enzyme inhibition or receptor antagonism). Unless wehave a clear idea of how an agent works, and possess specific in vitro assays to measure pharmacological activity (along with corresponding animal models), then establishing the role of the carbohydrate funtionality will be exceptionally demanding. However, the mainly phenomenological evidence described in this article does suggest that introduction of carbohydrate character into low molecular weight molecules can influence solubility (doxorubicin, elsamicin), pharmacokinetics (digoxin) and membrane transport (bleomycin), but this information is insufficiently strong to allow its prospective use in the design of novel agents.

As the sophistication of carbohydrate chemistry grows, newer agents will provide a suitable vehicle for establishing more sophisticated structure-activity relationships, and as a result generate derivative molecules with desirable characteristics including improved potency, selectivity and toxicity. These advances whether driven by low molecular weight or high molecular weight glyconjugates will almost certainly result in the discovery of new agents with novel and chemically useful pharmacologies.

ACKNOWLEDGEMENTS

We thank Katharine Irvine for excellent secretarial assistance.

REFERENCES

(1) Knight P (1989). The carbohydrate frontier. Biotechnology 7:35-40.
(2) Loscalzo J, Braunwald E (1988). Tissue plasminogen activator. New England J. Med. 319:925-931.
(3) Itri LM, Campion M, Dennin RA, Palleroni AV, Gutterman JU, Groopman JE, Trown PW (1987). Incidence and clinical significance of neutralizing antibodies in patients receiving recombinant interferon alpha 2a by intramuscular injection. Cancer 59:668-674.
(4) Von Wussow P, Freund M, Block B, Diedrich H, Poliwoda H, Deicher H (1987). Clinical significance of anti-IFN-α antibody titres during interferon therapy. The Lancet 635-636.
(5) Sneader W (1985). "Drug Discovery. The Evolution of Modern Medicines". New York: John Wiley and Sons.
(6) Fulton JF (1934). Charles Darwin (1758-1778) and the history of the early use of digitalis. Bull NY Acad Med 10:496-506.
(7) Estes JW, White PD (1968). William Withering and the purple foxglove. Scientific American 110-119.
(8) Withering W (1785). "An Account of the Foxglove and some of its Medicinal Uses." Birmingham: G.G.J. and J. Robinson.
(9) Friend DG (1934). Digitalis after two centuries. Arch Surg 111:14-19.

(10) Ackernecht, E (1962). Aspects of the history of therapeutics II. Digitalis and some other panaceas. Bull Hist Med 36:389-419.
(11) Paterson L (1967). The history of cardiac glycosides. Applied Therapeutics 9:60-65.
(12) Opie LH, Chatterjee K, Gersh BJ, Harrison DC, Kaplan NM, Marcus FI, Singh BN, Sonnenblick EH, Thadani U (1987). "Drugs for the Heart." New York: Grune and Stratton, p 91-110.
(13) Henderson FG, Chen KK (1962). Cardiac activity of newer digitalis glycosides and aglycones. J Med Chem 5:988-955.
(14) Brown BT, Stafford A, Wright SE (1962). Chemical structure and pharmacological activity of some derivatives of digitoxigenin and digoxigenin. Brit J Pharmacol 18:311-324.
(15) Elderfield RC (1935). The chemistry of the cardiac glycosides. Chem Reviews 17:187-249.
(16) Jaques LB (1978). The discovery of heparin. Sem Thrombosis Hemostasis 4:350-353.
(17) Best CH (1959). Preparation of heparin and its use in the first clinical cases. Circulation 19:79-86.
(18) Scrip (1988). Sanofi's rights issue and future strategy. 1371:12
(19) Scrip (1988). KabiVitrum's Fragmin launched in Sweden. 11296:24
(20) Dorr RT, Fritz WL (1980). "Cancer Chemotherapy Handbook." New York: Elsevier.
(21) Henry DW (1979). Structure activity relationship among daunorubicin and adriamycin analogs. Cancer Treatment Reports 63:845-854.
(22) DiMarco A, Casazza AM, Dasdia T, Necco A, Pratesi G, Rivolta P, Velcich A, Zaccara A, Zanino F (1977). Changes of activity of daunorubicin, adriamycin and stereoisomers following the introduction or removal of hydroxyl groups in the amino sugar moiety. Chem Biol Interactions 19:291-302.
(23) Carter SK, Crooke ST, Umezawa, H (1978). "Bleomycin: Current Status and New Developments." New York: Academic Press.
(24) Aoyagi Y, Katauo K, Suguna H, Primeau J, Chang L-H, Hecht S (1982). Total synthesis of bleomycin. J Amer Chem Soc 104:5537-5538.
(25) Umezawa H (1976). Structure and action of bleomycin. Prog Biochem Pharmacol 11:18-27.

(26) Bakhaeva GP, Berlin YA, Boldyreva EF, Chuprunova OA, Kolosov MN, Soifer VS, Vasiljeva TE, Yartseva IV (1968). The structure of aureolic acid (mithramycin). Tetrahedron Letters 32:3395-3598.
(27) Grundy WE, Glodstein AW, Rickher CH, Hanes ME, Warren HB, Sylvester JC (1953). Aureolic acid, a new antibiotic. I. Microbial studies. Antibiotics and Chemotherapy 3:1215-1217.
(28) Rao KV, Cullen WP, Sobin BA (1962). A new antibiotic with antitumour properties. Antibiot Chemother 12:182-186.
(29) Ryan WA, Schwartz TB, Perlia CP (1969). Effects of mithramycin on Paget's disease of bone. Ann Intern Med 70:549-557.
(30) Arnold AM, Whitehouse JMA (1981). Etoposide: a new anti-cancer agent. The Lancet 2:912-915.
(31) EORTC (1973). Epipodophyllotoxin VP 16213 in treatment of acute leukaemia, haematosarcoma, and in solid tumours. Brit Med J 3:199-202.
(32) Waksman SA (1958). "My Life With the Microbes." London: Robert Hale.
(33) Waksman SA (1951). Streptomycin isolation, properties and utilization. J Hist Med 6:318-347.
(34) Lietman PJ (1985). Aminoglycosides and spectinomycin:aminocyclitols. In Mandell GL, Douglas RG, Bennett JE (eds): "Anti-Infective Therapy," New York: John Wiley and Sons, p 109.
(35) Schatz A, Bugie E, Waksman SA (1944) Streptomycin, a substance exhibiting antibiotic activity against gram positive and gram negative bacteria. Proc Soc Exp Biol Med 55:66-69.
(36) Steigbigel NH (1985). Erythromycin, lincomycin and clindamycin. In Mandell GL, Douglas RG, Bennett JE (eds): "Anti-Infective Therapy," New York: John Wiley and Sons, p 197.
(37) Dutcher J (1968). The discovery and development of amphotericin B. Dis Chest 54 suppl:296-298.
(38) Cybulska B, Jakobs E, Falkowski L, Borowski E (1975). Structure-selective toxicity relationship in polyene macrolide antifungal antibiotics. In Lyr H, Polter C (eds): "Systemic Fungicides," Berlin: Akademie Verlag.
(39) Konishi M, Sugawara K, Kofu F, Nishiyama Y, Tomita K, Miyaki T, Kawaguchi H (1986). Elsamycins, new tumor antibiotics related to chartreusin I. Production, isolation, characterization and antitumor activity. J. Antibiotics XXXIX:784-791.

(40) Tamaoki T, Nomoto H, Takahashi I, Kato Y, Morimoto M, Tomita F (1986). Staurosporine: a potent inhibitor of phospholipid/calcium-dependent protein kinase. Biochem Biophys Res Comm 135:397-402.
(41) Smith CD, Glickman JF, Chang K-J (1988). The antiproliferative effects of staurosporine are not exclusively mediated by inhibition of protein kinase C. Biochem Biophys Res Comm 156:1250-1256.
(42) Nakanishi S, Matsuda Y, Iwahashi K, Kase, H (1986). K252b c and d: potent inhibitors of protein kinase C from microbial origin. J Antibiotics XXXIX:1066-1071.
(43) Kase H, Iwahashi K, Matsuda Y (1986). K252a: a potent inhibitor of protein kinase C from microbial origin. J Antibiotics XXXIX:1059-1065.

ROLE OF GALAPTIN IN OVARIAN CARCINOMA ADHESION TO EXTRACELLULAR MATRIX IN VITRO[1]

Howard J. Allen[2], Daniel Sucato[2], Barbara Woynarowska[3], Sally Gottstine[2], Ashu Sharma[2], and Ralph J. Bernacki[3]

Departments of Surgical Oncology[2] and Experimental Therapeutics[3], Roswell Park Memorial Institute, Buffalo, NY 14263

The binding of specific cell surface receptors by extracellular matrix components and the specific interaction of matrix components with each other are presumed to play critical roles in cell adhesion and motility. Perturbation of these interactions in malignancy is believed to facilitate the early stages of the metastatic process, specifically loss of normal adhesions and the development of matrix invasion.

There is considerable evidence that protein-protein interactions are involved in cell-matrix interactions. However, the vast data on altered cell surface glycoconjugates of transformed cells (1,2) and on endogenous lectins (3,4,5) drives the hypothesis that these glycoconjugates and lectins play a role in metastasis via modulation of cell adhesion. One carbohydrate-binding transformation-sensitive putative adhesion molecule is the differentiation-dependent β-galactoside-binding lectin sometimes referred to as galaptin.

The studies reported here were initiated to assess the role of galaptin in adhesion of ovarian carcinoma cells to extracellular matrix (ECM) in vitro. The ECM used in these studies was that synthesized in vitro by bovine corneal endothelial cells (BCEC). Human ovarian carcinoma exhibits interesting characteristics which include metastatic growth on the peritoneum, shedding of cells into the peritoneal

---

[1]This work was supported by NIH grants Ca 42584 (H.J.A.), CA 42898 (R.J.B.) and Institutional Research Grant IN 54-28 of the American Cancer Society.

cavity, growth of multicellular clusters in vivo, apparent re-adherence of cells to the peritoneum, and synthesis and release of lactosaminoglycans in vitro. Immunohistochemical studies indicated that galaptin is a major protein of ovarian carcinoma cells present in patient effusions and it is distributed throughout the cytoplasm. Enzyme-linked immunoadsorbent assay (ELISA) and immunoprecipitation experiments demonstrated that galaptin is also a major protein of the A121 ovarian carcinoma cell line, constituting $\leq$1% of extractable protein bound by DEAE Sephacel. Western blot analyses revealed that the galaptin present in ovarian carcinoma consists of a 14.5 kDa subunit. A121 cells also display surface receptors for galaptin: binding sites/cell = 3×10^8 and $K_a = 1.2 \times 10^9$ M^{-1}. The presence of galaptin in BCEC and BCEC-derived ECM was demonstrated by ELISA. Of the total ECM-bound immunoreactive galaptin, about 75% appears to be insoluble in phoshate buffered saline-lactose. ECM was also found to contain receptors for galaptin. Treatment of ECM with lactose increased the apparent receptor density: binding sites/cm^2 = 7×10^{13} and $K_a = 2.6 \times 10^9$ M^{-1}. Pretreatment of A121 cells with galaptin inhibited adhesion to ECM. The addition of exogenous galaptin to ECM had slight, but variable effect, on cell adhesion. The data suggest that early adhesion events may be carbohydrate-specific involving interaction between ECM-bound galaptin and cell surface galaptin receptors.

REFERENCES

1. Hakomori SI (1985). Aberrant glycosylation in cancer cell membranes as focused on glycolipids: Overview and perspectives. Canc Res 45:2405.
2. Dennis JW (1988). Asn-linked oligosaccharide processing and malignant potential. Canc Surv 7:573.
3. Allen HJ, Karakousis C, Piver MS, et al (1987). Galactoside-binding lectin in human tissues. Tumor Biol 8:218.
4. Lotan R, Raz A (1988). Endogenous galactoside-specific lectins as mediators of tumor cell adhesion. J Cellul Biochem 37:107.
5. Monsigny M, Roche A-C, Kieda C, Midoux P, Obrenovitch A (1988). Characterization and biological implications of membrane lectins in tumor, lymphoid and myeloid cells. Biochimie 70:1633.

ANALYSIS OF N-GLYCOSYLATION MUTANTS IN DICTYOSTELIUM DISCOIDEUM

Freeze HH, Koza-Taylor P, Jones, JA, and Loomis WF

La Jolla Cancer Research Foundation
La Jolla, California 92037

N-linked oligosaccharides of *Dictyostelium discoideum* are derived from the usual lipid-linked oligosaccharide (LLO) precursor, $Glc_3Man_9GlcNAc_2$-P-P-dolichol. Mutant strains HL241 and HL243 lack an antigenic determinant (CA1) found on sulfated N-linked oligosaccharides because they synthesize a truncated LLO, $Man_6GlcNAc_2$-P-P-dolichol, due to the apparent lack of mannosyl transferase(s)(1-3). Here we report on the consequences of this alteration in LLO synthesis and structure for both *in vivo* and cell-free N-glycosylation.

Cell-free N-glycosylation was measured by transfer of endogenous LLO to octanoyl-Asn-[^{125}I] Tyr-ThrNH$_2$(4). This assay showed that the rate and extent of glycosylation was 5-10x lower in the mutants than in the wild-type. This was not due to differential stability of the system, or to limited availability of acceptor peptide or its Km. The results suggested that the LLO might be rate limiting. However, when intact cells were metabolically labeled with ^{3}H-Man, the rates of synthesis and turnover and the relative amount of LLO in each strain were nearly the same. These results suggest that the amount of LLO actually available for transfer within the microenvironment of the oligosaccharyl transferase may be limiting. This may not be surprising since the *in vitro* assay depends on a relatively static and non-renewable supply of endogenous of LLO.

This work was supported by NIGMS32485.
H.F. is an Established Investigator of the American Heart Association.

In contrast, when glycosylation was measured in intact cells using [^{3}H]GlcN and Man, we found that strain HL241 was nearly normal while HL243 was 3-4 fold lower than the wild-type. We interpret this to mean that in living cells the truncated LLO(HL241) can be transferred to protein as efficiently as the normal LLO in wild-type cells. This may be because the system is dynamic and self-replenishing which is not the case in the *in vitro* assay. The decrease in glycosylation seen only in HL243 may be a consequence of the pleiotropic effects of the primary mutation rather than a direct result of the altered LLO structure. Furthermore, these results suggested that although the two strains share a common lesion in the expression of CA1 and LLO biosynthesis, they must differ in other ways. We, therefore, conducted a genetic analysis of the strains.

HL241 can complete development while HL243 cannot even initiate development. When HL241 is crossed with a wild-type strain which expresses CA1, the diploid strain expresses CA1; therefore, the mutation in HL241 is recessive. When HL243 is crossed with the wild-type (or with HL241), CA1 is not expressed, so the mutation in HL243 is dominant. Moreover, haploid segregants of this diploid strain co-segregate CA1 and developmental competence. Since dominant mutations are rare, it is likely that the same dominant mutation in HL243 is responsible for the loss of CA1 and the failure to develop. The results further suggest that the mutations in the two strains do not reside in the same locus.

REFERENCES

1. Knecht DA, Diamond RL, Wheeler S, Loomis WF: J Biol Chem 259:10633-10640, 1984.
2. Freeze HH, Mierendorf R, Wunderlich R, Dimond, RL: J Biol Chem 259:10641-10643, 1984.
3. Freeze HH, Willies L, Hamilton S, Koza-Taylor P: J Biol Chem 264:5653-5659, 1989.
4. Wieland FT, Gleason ML, Serafini TA, Rothman JE: Cell 50:289-300, 1987.

THE ROLE OF GLYCOSYLATION IN THE TRANSPORT OF RECOMBINANT GLYCOPROTEINS THROUGH THE SECRETORY PATHWAY OF LEPIDOPTERAN INSECT CELLS[1].

Donald L. Jarvis, Christian Oker-Blom[2], and Max D. Summers

Department of Entomology and Institute of Biosciences and Technology, Texas A&M University, College Station, Texas 77843

Baculovirus expression vectors have become extremely important tools for the high level expression of foreign gene products in a eucaryotic host (1,2). The most commonly used hosts for baculovirus-mediated foreign gene expression are cell lines derived from Lepidopteran insects. These cells appear to carry out most of the protein processing events that take place in vertebrate cells (1,2). However, the protein processing pathways in these cells have not been studied at the molecular level. With widespread and growing use of the Baculovirus expression system, we must obtain a much better understanding of the protein processing pathways in Lepidopteran insect cells. In this study, we examined the role of N-glycosylation in the transport of foreign proteins through the secretory pathway of Sf9 cells.

Previously, we constructed a recombinant baculovirus that would express human tissue plasminogen activator (t-PA) in infected Sf9 cells (3). The recombinant t-PA polypeptide was N-glycosylated in these cells and it underwent secretion. t-PA secretion was prevented by treatment of the cells with tunicamycin (TM), which blocks the covalent addition of N-linked oligosaccharides. However, t-PA secretion was unaffected by treatment with castanospermine (CS), an inhibitor that blocks the first step in the processing of N-linked oligosaccharides. CS effectively blocked processing by the insect cell glucosidase, as shown by an increase in the molecular weight of t-PA. These results suggested that N-glycosylation per se, but not oligosaccharide processing, is required for the secretion of recombinant t-PA from baculovirus-infected Sf9 cells. This posed two major questions for further study. What are the roles of N-glycosylation and oligosaccharide processing in the movement of recombinant glycoproteins, in general, through the Sf9 cell secretory pathway? What

[1]This work was supported by funds from the Institute of Biosciences and Technology, Texas A&M University.
[2]Present Address: Labsystems Oy; Pulttitie 8, Helsinki, Finland.

molecular mechanism explains how the secretion of some proteins is blocked in TM-treated Sf9 cells?

We have addressed both of these questions by surveying the effects of TM or CS treatment on the secretion or cell surface localization of several different recombinant glycoproteins expressed in baculovirus-infected Sf9 cells. Unlike the results obtained with t-PA, TM did not block the secretion of a nonglycosylated form of human ß-interferon. TM blocked the appearance of the structural glycoproteins of Sindbis virus on the surface of Sf9 cells. However, cell surface expression of a baculovirus envelope glycoprotein was not blocked by TM. CS increased the size of all glycoproteins tested, but it did not block the secretion or cell surface localization of any. These results showed that N-glycosylation is required for the transport of some, but not all, foreign glycoproteins through the Sf9 cell secretory pathway, while oligosaccharide processing generally appears to be dispensable.

We also found that the nonglycosylated precursors of several different recombinant glycoproteins were tightly associated with at least two other proteins in TM-treated Sf9 cells, designated p80 and p31. These proteins were not antigenically related to any of the recombinant products, as shown by western blotting. Their presence was always inversely correlated with transport through the secretory pathway. We propose that a physical interaction between p80 and/or p31 and a nonglycosylated recombinant glycoprotein precursor in TM-treated Sf9 cells produces a transport-incompetent heterooligomeric complex. If this speculation is correct, then p80 and/or p31 would be insect cell analogs of the mammalian immunoglobulin heavy chain binding/glucose-regulated 78 kilodalton protein (BiP/GRP78), for which this activity has been proposed previously (4; reviewed in 5).

References

1. Luckow VL, Summers MD (1988). Trends in the development of baculovirus expression vectors. Bio/Technology 6:47.
2. Miller LK (1988). Baculoviruses as gene expression vectors. Ann Rev Microbiol 42:177.
3. Jarvis DL, Summers MD (1989). Glycosylation and secretion of human tissue plasminogen activator in recombinant baculovirus-infected insect cells. Mol Cell Biol 9:214.
4. Haas IG, Wabl M (1983). Immunoglobulin heavy-chain binding protein. Nature 306:387.
5. Pelham HRB (1986). Speculations on the functions of the major heat shock and glucose-regulated proteins. Cell 46:959.

Index

Acetylcholinesterase (AChE)
    anchor structure. *See* Glycoinositol phospholipid anchors
    fragment analysis, 7–8
    glycoinositol phospholipid anchor, 1
        resistance to PIPLC, detection by anchor acylation, 11–12
        structure of, 6–10
    glycoinositol phospholipid anchor, identification of, 3–5
    histochemical detection of, 11
    hydrophobic domain, identification of, 3
    membrane anchor components, identification of, 4–5
    membrane–bound, classes of, 2
    purification of, 3
ß-N-Acetylgalactosaminidase, sperm receptor activity and, 63–64
N-Acetylgalactosaminyltransferase, 107
N-Acetylglucosamine (GlcNAc), 108, 146, 147
ß-N-Acetylglucosaminidase, sperm receptor activity and, 63–64
AChE. *See* Acetylcholinesterase
Activase, 175
Adenylate cyclase, 12
Adriamycin, 173
Aedes albopictus cells, glycosylation characteristics of, 161
AIDS, 174
Alkaline phosphatase, 128
Alkylacylglycerol, 10
1-Alkyl-2-acylglycerol, 1
Alzheimer's disease, 37, 49
Aminopeptidase N, 128
Amphotericin, 185–186
Amyloidosis, murine systemic, 37
Amyloids, 40, 45
Amyotrophic lateral sclerosis, 49
Anchor substitution, 21–22
Anemia of renal insufficiency, 174
Anticoagulant therapy, 173, 178–180
Antiinfectives, 184–190
Antineoplastics, 180–184
Attachment, 60–63

Baculovirus expression system, 169, 199
Binding, 62
Binding/glucose-regulated 78 kilodalton protein (BiP/GRP78), 200
Bleomycin, 181–182, 189
Bovine spongiform degeneration (BSE), 36, 47

Calcium, PN HEV attachment and, 76, 77
Cancer therapy
    with Bleomycin, 181–182
    with Doxorubicin, 180–181
    with Etoposide, 183–184
    with Mithramycin, 182–183
Carbohydrate recognition domains (CRDs), 81
Carbohydrates
    in adhesive recognition, 75–77
    analysis method, 162
    of GCR, processing by Sf9 cells, 170
    HEV and, 77–78
    in lymphocyte-HEV interactions, 99–100
    as recognition determinants in homing specificities, 84
    in sperm receptor activity, 62
    structure–activity relationships, pharmacological activity and, 174–176, 189
Castanospermine, 159, 167, 199–200
CD8, 19–22
CD44 antigen, 97
Choriocarcinoma, diagnosis of, 115, 121–124
Choriocarcinoma hCG
    N-linked sugar chains, structural alterations in, 117–119
    O-linked sugar chains, structural alterations in, 119–121
Chorionic gonadotropin, human. *See* Human chorionic gonadotropin (hCG)
Chronic wasting disease, 36, 47
CJD. *See* Creutzfeldt-Jakob disease
Colony stimulating factors, 174
Congestive heart failure, 173

Index

CRDs (carbohydrate recognition domains), 81
Creutzfeldt-Jakob disease
  amyloids in, 40
  familial, 46, 49
  inheritence of, 46
  protease-resistant prion proteins in, 35, 36, 40
  PrP gene and, 37
  sporadic, 46
Cyclic AMP phosphodiesterase, 12
Cytoplasmic extensions, 18–20

Decay-accelerating factor (DAF), 13, 17, 18
Deoxymannojirimycin, 159, 166
Deoxynojirimycin, 159
Detergent-lipid protein complexes (DLPC), 35
Dictyostelium discoideum, N-glycosylation mutants, 197–198
Differentiation
  EC cell, 111
Differentiation, cell, 128–129
Digitalis lanata, 176
Digitalis purpurea, 176, 178
Digoxin, 176–178, 189
Dimyristoylphosphatidylinositol, 6, 9
Diolein, 32
Dipeptidylpeptidase IV, 128
Diradylglycerol, 9
DLPC (detergent-lipid protein complexes), 35
Doxorubicin, 180–181, 189
Drug discovery, 173

EC cell differentiation, 111
Egg-binding proteins, 59–60, 66
ELAM-1, 70, 84, 98, 99
Elsamicin A, 186–187
Elsamicin B, 186–187
Encephalopathy, anoxic, 37
Endothelial cell adhesion molecules, 98–99
Enterocyte differentiation, 127–129, 132–141
Epidermal growth factor domain, of gp90^{MEL-14}, 82–83, 84
Erythromycin, 173, 185
Erythropoietin, 173, 174
Escherichia coli, glycosylation by, 174
Etoposide (VP-16), 173, 183–184
Excreted factor, 32; *see also* Lipophosphoglycan

Exoglycosidases, effect on sperm receptor activity, 62–64
Extracellular matrix adhesion of ovarian carcinoma, galaptin and, 195–196

Fertilization, sperm-egg interaction during, 59–60
Flow cytometric analyses, 19
Fructose-1-phosphate, 75
Fucoidin, 76
α-Fucosidase, sperm receptor activity and, 63–64

Galactose, 66
α-Galactosidase, sperm receptor activity and, 63–64
ß-Galactosidase, sperm receptor activity and, 63–64
ß1,4 Galactosyltransferase. *See* GalTase
Galactose oxidase, sperm receptor activity and, 64–65
Galaptin, in ovarian carcinoma adhesion to extracellular matrix *in vitro*, 195–196
GalTase
  cell surface form, 107
    differential regulation of, 110–112
    properties of, 108
  functions of, 107–108
  Golgi form, 107
    differential regulation of, 110–112
    properties of, 108
  localization of, 107
  molecular biology of, 109–110
  multiple protein structures for, 109
Gaucher's disease, 159–161
GCR. *See* Glucocerebrosidase
Gerstmann–Sträussler syndrome (GSS)
  amyloids in, 40
  delayed onset of, 46
  inheritance of, 46
  protease-resistant prion proteins in, 35, 36, 40
  PrP gene and, 37, 48
GlcNAc (N-acetylglucosamine), 108, 146, 147
Glucocerebrosidase
  expression in *Spodoptera frugiperda* Sf9 cells, 159
  in Gaucher's disease, 160
  recombinant
    characterization of, 164–165
    effects of oligosaccharide processing inhibitors, 166–168

expression of, 164
glycosylation of, 165–166
oligosaccharide structural characterization, 168–169
signal sequence, cleavage of, 171
vector constructions, 162–163
ß-Glucuronidase, sperm receptor activity and, 63–64
N-Glycan processing, 127, 129, 139–141
Glycine, in glycoinositol phospholipid anchor, 5
Glycoconjugates, 173, 174; see also specific glycoconjugates
Glycoinositol phospholipid anchors
analysis of, procedures for fragment generation for, 7
components of, 5
identification in human erythrocyte AChE, 3–5
location of, 2
mRNA transcription, 17
processing of, 17–18
regulation, by susceptibility to phospholipase cleavage, 12–13
structure of, 6–10
Glycoinositol phospholipids
nonionic detergent–binding domain of, 11
susceptibility to phospholipase C cleavage, 13
Glycolipid anchor. See Glycoinositol phospholipid anchors
Glycophospholipid anchor
processing, 20
Glycoproteins
in nuclear pore, identification of, 146–147
structure-activity relationships, 175
transport of, glycosylation and, 199–200
Glycosylation
of hCG, in trophoblastic diseases, 115–125
in transport of glycoproteins, 199–200
N-Glycosylation, 197
N-Glycosylation mutants, in *dictyostelium discoideum*, 197–198
Glycosyl-phosphatidylinositol
Glycosyl phosphatidylinositol anchors, 35–36
in prion proteins, 40
Glycosyltransferases. See GalTase
functions of, 107
Golgi

GMP–140, 70, 84
gp90^{Hermes}
biochemical characterization of, 95–97
and lymphocyte-HEV interactions, 96
molecular cloning of, 95–97
mucosal addressin and, 91, 100–101
as novel cell adhesion molecule, 97–98
gp90^{MEL-14}
HEV-ligand for, 83
as lectin, confirmation of, 80–83
lectin domain of, 81–83
lectin-like receptor and, 78–80
molecular cloning of, 69
molecular organization, 84
PPME-bead binding and, 79
GSS. See Gerstmannn–Sträussler syndrome

Hansenula holstii, 76
hCG. See Human chorionic gonadotropin
HEBF$_{PN}$, 80
Heparin, 173, 178–180
Hermes-3 antigen, 101
Hermes homing receptor, 101
Heterologous expression, 169–170
High endothelial venules (HEV)
carbohydrates and, 77–78
interactions with lymphocytes, 74, 92–93
lymphocyte migration, 69, 71, 72
lymphocyte migration and, 91, 92
Histidine, in glycoinositol phospholipid anchor, 5
Homing, 69
Homing receptors
calcium-dependent, lectin-like, 77, 80
classes of, 74
definition by monoclonal antibodies, 93–95
function of, 69
HT-29 cells
cell differentiation and, 128–129
culture conditions for, 130
differentiated, mannosidase I, 138
growth curves and sucrase specific activity, 132–134
undifferentiated, 127
accumulation of mannose species, 134
alteration of high mannose processing, 134–138
degradation of high mannose glycopeptides, 139
mannosidase I, 138

Index

Human chorionic gonadotropin (hCG)
  detection of, 115
  malignant vs. normal, detection of, 115
  production of, 115–116
  structural alterations
    diagnosis of choriocarcinoma and, 121–125
    in N-linked sugar chains, 117–119
    in O-linked sugar chains, 119–121
  structural characterization of, 115–116
  sugar moieties of, 115–116
Huntington's disease, 46

ICAM–1, 98–99
Infections, 173, 174
Inositol, palmitoylation of, 10
Inositol phospholipid, in human erythrocyte AChE, structure of, 9, 10
Insect cells, 161, 199–200
Interferons, 174
Invasive mole, sugar chains of hCG in, 118

Kala-azar, 23
Kozak consensus sequence, 163
K252 series of antibiotics, 188
Kuru
  amyloids in, 40
  cannibalism and, 36
  protease-resistant prion proteins in, 35, 40
  spread of, 45–46

LEC-CAM family, 70, 84–85
Lectin-like receptor, relationship to gp90[MEL-14], 78–80
Lectins, calcium-dependent, 80
Leishmania donovani
  cells and cell culture method, 24
  cloning of, 24–25
  lipophosphoglycan
    attenuation of oxidative burst in monocytes, 28, 29
    inhibitory effect on protein kinase C, 28, 30–31
    metabolic labeling and extraction in characterization of RT5 line, 26–28
    procedure, 25
  mutagenesis of, 24–25
  RT5 ricin-resistant clone
    characterization of, 26
    infection of monocytes, 26–28
    selection of, 26
  survival in hostile environment, 23–24
  wild type, in human monocytes, 26–28
Lepidopteran insect cells, transport of recombinant glycoproteins, 199–200
LFA-1, 74, 95
LFA-3, 13
Limax agglutinin, 77–78, 84
Lipid-linked oligosaccharide precursor (LLO), 197–198
Lipophosphoglycan (LPG)
  calcium chelation, 32–33
  inhibitory effect on protein kinase C, 28, 30–31
  luminol reaction and, 31
  1-O-alkylglycerol portion of, 32, 33
  oxidative burst and, 28, 29, 31–32
  phosphosaccharidyl-PI fragment of, 28, 30–31
  structure of, 24
  survival of leishmanial parasites and, 23, 24, 31
LPAM-1, 95
LPG. *See* Lipophosphoglycan
Lymphocyte-endothelial adherence, specificity of, 72–74
Lymphocyte-HEV interactions
  carbohydrates and, 99–100
  gp90[Hermes] and, 96
Lymphocyte homing, 69
Lymphocyte homing receptors. *See* Homing receptors
  vascular addressins and, 91–101
Lymphocyte recirculation, 70–72, 92
Lymphocytes
  carbohydrate-binding receptor of, 75–77
  interactions with high endothelial venules, 92–93
Lymphomas, HEV-binding, 100–101
Lysosomal targeting, 160–161, 169–170

Man_{9-8}-GlcLNAc$_2$, 127–128, 129
Man/GlcNAc receptor, 127–129
(^{3}H-)Mannose, conversion of, 131–132
Mannose-6-phosphate (M6P), 75
Mannosidase I, 127, 138
MECA-367, 101
MEL-14 antibody, 69, 79–80
MEL-14 antigen, 94–95, 100
Metabolic cell labeling method, 130–131, 162
N-Methyldeoxynojirimycin, 159
N-Methyl-N-nitroso-N'-nitroguanidine, 24–25

Mithramycin, 182–183
Monoclonal antibodies
  definition of homing receptors, 93–95
  identification of vascular addressins, 98
  against PrP 27–30, 35
Monocytes, human
  attenuation of oxidative burst by LPG, 28, 29
  isolation of, 25–26
Mouse egg, interaction with sperm, 59–67
Mucosal addressin (MAd)
  biochemical features of, 99, 100
  functions of, 100–101
  identification of, 98
Murine Qa-2, 18
Mycobacterium tuberculosis, 184
Myelosuppression, 174
Myocardial infarction, 174, 175

Neuraminidase, sperm receptor activity and, 63–64
Nuclear envelope, structure of, 145–146
Nuclear localization sequences, microinjection assay for, 149–150
Nuclear membrane, structure of, 145–146
Nuclear pore
  in nuclear protein uptake, 147–150
  in RNA transport, 151–152
Nuclear pore glycoproteins
  identification of, 146–147
  molecular cloning of, 152–154
Nuclear pore protein p62
  domain structure of, 154
  isolation and sequencing of, 152–154
Nuclear protein transport, *in vitro* assay for, 150
Nuclear protein uptake, nuclear pore in, 147–150
Nuclear transport, 150

Oligosaccharides, ZP3-derived O-linked, 60, 62
O-linked GlcNAc, 146–147
Ovarian carcinoma, adhesion to extracellular matrix, 195–196
Oxidative burst, 28, 29

Palmitate, in glycoinositol phospholipid anchor, 5
Palmitoylated inositol, 9–10
Palmitoylation, of inositol, 12
Papain, 4–5
Parkinson's disease, 49

Pasturella pestis, 184
Peripheral lymph node addressin (PNAd), 98, 99
Peripheral lymph nodes, 92
Peyer's patches, 70, 72, 92
Pgp-1 antigen, 97
Phorbol myristic acid (PMA)
  PN HEV attachment and, 76–77
  PPME-binding and, 76–77
  stimulation of oxidative burst, 28, 31
Phosphatidylinositol-specific phospholipase C (PIPLC)
  AChE release and, 6, 7
  protein release and, 1, 2–3
  resistance of PrP^{Sc}, 42
  resistance to, 1–2, 10
Phosphatidylserine, 28, 32
Phospholipase C, 12
Phospholipid methyltransferase, 12
Phycoerythrin, 155
PIPLC. *See* Phosphatidylinositol-specific phospholipase C
Plasmanylinositol, 9
Plasmid constructions, 18–19
PMA. *See* Phorbol myristic acid
PN HEV attachment, PMA and, 76–77
Podophyllotoxin, 183
Polyhedrin promoter, 162, 163
Polyhedrin regulatory sequences, 163
PPME, 76
PPME-bead binding, $gp90^{MEL-14}$ expression and, 79
Prion diseases
  of animals, 35, 36
  of humans, 35, 36
  infectious, sporadic and genetic, 45–47
  PrP^C in, 43
Prion proteins (PrP)
  isoforms of, 35, 39–40
  post-translational modifications, 41–42
  protease-resistant, 40
Prions
  abnormal PrP isoforms and, 35
  brain degeneration and, 36
  multiple forms of, 43–45
  novel and unprecedented nature of, 47–49
Pronase, 9
Protein kinase A, 32
Protein kinase C
  induction of oxidative burst, 32
  inhibition by LPG, 28, 30–31, 32
Protein kinase M, 32

Index

Proteins, with glycoinositol phospholipid anchors, 2–3
PrP. *See* Prion proteins
PrP 27–30, 36–38
PrP 27–30 antibodies, 45
PrP 39–40, 42–43
PrP^{CJD}, in animal and human prion disease, 47–48
PrP genes
   in mice, 37
   open reading frame of, 38–39
   structure and organization of, 38–39
PrP isoform, 35, 39–40
PrP^{Sc}
   in animal and human prion disease, 47–48
   polymerization, 39–40
   in prions, 36–38
Pyruvate dehydrogenase, 12

Quabain, 178

Receptor oligosaccharides, 59
Recirculation, lymphocyte, 70–72, 92
Recombinant virus production, 161–162
Reductive readiomethylation, 4, 5
Rheumatoid arthritis, 73
RNA transport, 145
Rous sarcoma virus, 42

Scrapie
   infectivity of, 48
   prions and, 35
   PrP^C in, 42
   transmission of, 47
Scrapie prion, 36–38
Scrapie PrP isoform (PrP^{Sc}), 35
Secretion, 159
Sf9 cells, 199–200
Sialic acid, 77–78
Sialytransferase, 107
S49 lymphoma cells, 76
Sperm-egg binding competition assay, *in vitro*, 60–62
Sperm receptors
   exoglycosidases and, 62–64
   galalctose oxidase and, 64–65
   loss of activity, 62
   of mice, 59
   mouse. *See* ZP3 sperm receptor
   recognition determinants, 66
Spodoptera frugiperda cells, 159, 169, 199–200

Stamper–Woodruff *in vitro* adherence assay, 72, 73, 75, 92
Staurosporine, 187, 188
Streptomyces argillaceus, 182
Streptomyces erythreus, 185
Streptomyces griseus, 184
Streptomyces nodosus, 185
Streptomyces peucetius, 180
Streptomyces vericillus, 181
Streptomycin, 173, 184–185, 185
Sucrase–isomaltase, 128
Sugar chains
   N-Linked, of choriocarcinoma hCG, 117–119
   O-linked, of choriocarcinoma hCG, 119–121
SV40 T antigen localization sequence, 147–150, 155
Swainsonine, 159, 166, 167

[^{125}I] TID labeling, 4–5
Tissue plasminogen activator, 173, 174, 175, 199
T–lymphocyte CD4 receptor, 174
Transfection procedures, 19
Transmembrane domain, of gp90^{MEL-14}, 82
Transmissible mink encephalopathy, 47
Tunicamycin, 199–200

UDPgalactose (UDPGal), 108

Variant surface glycoproteins (VSGs), 6, 9–10, 13
Vascular addressins
   biochemical features of, 99–100
   identification by monoclonal antibodies, 98
   mucosal, 98–101
VP–16 (Etoposide), 173, 183–184
VSGs (variant surface glycoproteins), 6, 9–10, 13

WGA–ferritin, 146–147
Wheat germ agglutinin (WGA), 146, 151–152

Zona pellucida, 59, 60
ZP3-derived O-linked oligosaccharides, 64–65, 66
ZP3 sperm receptor
   biochemical characteristics, 59, 60
   galalctose oxidase and, 64–65
Zymosan, opsonized, 28, 29, 31